Botany: Theory and Concepts

Edited by **Harvey Parker**

New York

Published by Callisto Reference,
106 Park Avenue, Suite 200,
New York, NY 10016, USA
www.callistoreference.com

Botany: Theory and Concepts
Edited by Harvey Parker

International Standard Book Number: 978-1-63239-106-3 (Hardback)

The publisher's policy is to use permanent paper from mills that operate a sustainable forestry policy. Furthermore, the publisher ensures that the text paper and cover boards used have met acceptable environmental accreditation standards.

Trademark Notice: Registered trademark of products or corporate names are used only for explanation and identification without intent to infringe.

Printed in the United States of America.

Contents

Permissions

List of Contributors

Preface

I am honored to present to you this unique book which encompasses the most up-to-date data in the field. I was extremely pleased to get this opportunity of editing the work of experts from across the globe. I have also written papers in this field and researched the various aspects revolving around the progress of the discipline. I have tried to unify my knowledge along with that of stalwarts from every corner of the world, to produce a text which not only benefits the readers but also facilitates the growth of the field.

Botany is a field of biology that deals with the science of plants. This book discusses current issues of plant-environment interaction which would be appealing to environmental scientists. In the beginning, this book talks about flood tensions, resilience to drought and dehydration, toxic effects of chromium and lead on plants. Reading further, it discusses the facets of economic botany including reports on smut disease in few crops and features of plant extracts. The conclusion encompasses topical issues like morphogenesis and genetics in cotton fiber special cell, hybrid lethality in the Genus Nicotiana, pollen tube growth in Leucojum aestivum etc. This book targets researchers, students and professionals involved in the field of botany.

Finally, I would like to thank all the contributing authors for their valuable time and contributions. This book would not have been possible without their efforts. I would also like to thank my friends and family for their constant support.

Editor

Part 1

Adaptations and Responses to Environmental Extremes

Drought and Desiccation-Tolerance and Sensitivity in Plants

Tobias M. Ntuli

*Plant Germplasm Conservation Research,
School of Biological and Conservation Sciences,
University of KwaZulu-Natal, Durban,
Department of Life and Consumer Sciences,
University of South Africa, Florida, Johannesburg,
South Africa*

1. Introduction

As sessile organisms, plants encounter a plethora of stresses in their natural environment (reviewed by Janská *et al.*, 2010). They must withstand these stressors to survive. Stresses are abiotic or environmental and biotic. Environmental stressors include water, temperature, salt or salinity, light and metal ions. Water stress consists of both low – osmotic stress or water deficit: drought and drying or dehydration or desiccation (relative humidity [RH] < 100%) and high (RH = 100%) – flood and imbibition or rehydration - water stress. Salinity and water stress are closely related. Whereas high salinity or hyperosmolality is desiccating, low salinity and/or hypo-osmotic shock promotes hydration. Similarly, temperature stress comprises low – chilling/cold (0 °C > T° < 25 °C), freezing (-40 °C < T° ≤ 0 °C) and cooling (T° < original temperature [usually < -40 °]) and high – warming or thawing (T° > original temperature [usually > -196 or -80 °C]), heat (25 °C ≤ T° 50 °C) and heat shock (T° > 50 °C) – temperature stress as are light and metal ions stressors. Biotic stresses comprise micro-organisms – viruses, bacteria and fungi, insects and herbivores.

Desiccation-tolerance is not synonymous with drought tolerance (Alpert, 2005)! On one hand, desiccation-tolerance of an organism is defined as the ability of a living structure to survive drying to equilibrium with low (< 50%) RH and maintain low intracellular water concentrations (WCs). Drought tolerance (*sensu stricto*) is survival of low environmental water availability while maintaining high internal water contents (WCs), on the other. A drought-tolerant organism that is not desiccation-tolerant will die if it loses much of its water, whereas a desiccation-tolerant organism will survive under the same conditions. Thus, *desiccation-tolerance is one mechanism of drought tolerance*!

Desiccation-tolerance is generally understood to pertain to organisms that will survive dehydration to an overall WC equal to, or less than, 0.1 g (water) per g dry mass (g g^{-1}) (Berjak, 2006). Anhydrobiosis is commonly used synonymously with desiccation-tolerance in literature but Berjak (2006) cautions against this habit as anhydrobiosis implies complete absence of intracellular water which is not the case at WCs of around 0.1 g g^{-1}.

Intracellular WCs of ≤ 0.1 g g^{-1} are considered to represent levels at which macromolecules can no longer be surrounded by a water monolayer, thereby precluding enzymatic activity and, thus, all metabolism (Billi and Potts, 2002). Unique water properties pertain at such low WCs. In fact, water should be viewed not only as an intracellular medium but also as a structural component of macromolecules, such as proteins, which were water to be completely removed, are likely to undergo conformational changes (Billi and Potts, 2002). In solution, proteins are held to exclude small molecules from their immediate vicinity, thus being surrounded by solute-free water (Timasheff, 1982, Parsegian, 2002).

Upon extreme dehydration, but assuming a residual layer of water remains in close association with proteins, small solutes – which must include inorganic ions – are likely to perturb the residual water, where their localised effects could contribute to macromolecule denaturation (Berjak, 2006). This phenomenon may well impose one constraint on biological material being able to survive completely anhydrous conditions, thus imposing a limitation on absolute desiccation tolerance (Berjak, 2006).

2. Drought tolerance

Environmental variables, especially those affecting water vailability and temperature, are the major determinants of plant growth and development (reviewed by Janská *et al.*, 2010). Drought is undoubtedly one of the prime abiotic stresses (reviewed by Ashraf, 2010). Crop yield losses due to drought stress are considrable.

Mechanisms of drought tolerance occur at four levels (reviewed by Ashraf, 2010). They include genetic fectors which underlie morphological adaptations, physiological acclimation and cellular adjustments. Morphological adaptations consists of improved root length and thickness, thick and/or waxy coverings of leaves, lower leaf size and weight, higher green leaf area, delayed leaf senescence and smaller epithelial cells. Physiological acclimation comprises higher stomatal conductance and density, slower rates of transpiration, early and reduced asynchrony between male and female flowering and maturation and better production, assimilation, accumulation and partitioning of biomass and seed yield. Cellular adjustments entail higher chlorophyll content and particle numbers or harvest index, lower osmotic potential and mechanisms of desiccation-tolerance (see below).

3. Desiccation-tolerance

Desiccation-tolerence is rare but universal (Alpert, 2005)! There may be considerable commonality among the mechanisms and processes facilitating desiccation-tolerance across the spectrum of organisms that show this trait (Berjak, 2006).

3.1 Occurrence

Desiccation-tolerance occurs in organisms or life stages of species or taxa of higher – 'resurrection plants' and 'orthodox seeds' – and lower – mosses – plants, animals – nematodes and bdelloid rotifers and microorganisms - terrestrial micro-algae, lichens: symbionts of fungi (mycobiont) and algae or cyanobacteria (photobiont), bacteria and yeast (Berjak, 2006).

'Resurrection plants', so far known to be constituted of 330 species, have been described from nine pteridophyte and ten angiosperm families (Proctor and Pence, 2002). Their vegetative tissues are characterised by desiccation-tolerance. The plants concerned are able to remain viable despite considerable dehydration, resuming metabolic activity when water becomes available. Desiccation-tolerance in angiospermous resurrection plants, like orthodox seeds, is based on a spectrum of mechanisms that accompany drying (Illing *et al.*, 2005). Developing 'orthodox seeds' acquire the ability to tolerate desiccation relatively early, preceding the final developmental phase of maturation drying on the parent plant by some time (Bewley and Black, 1994; Vertucci and Farrant, 1995 ; Kermode and Finch-Savage, 2002).

Although desiccation-tolerance appears to be far more restricted across the spectrum of animals than plants (Alpert, 2005), the phenomenon has been documented for nematodes (Solomon *et al.*, 2000 ; Browne *et al.*, 2002) and bdelloid rotifers (Laprinski and Turnacliffe, 2003 ; Caprioli *et al.*, 2004), for example, and, classically, for the encysted embryos of the brine shrimp, *Artemia* spp. (Clegg, 1986, 2005).

Lichens represent a symbiosis between a fungus – the mycobiont and a green alga or cyanobacterium - the photobiont. The remarkable outcome of the symbiosis is that neither of the partners remains constrained to the cryptic habitats that would be obligatory for either one alone (Kranner *et al.*, 2005).

Prokaryotes in soils, which might periodically become very dry, need to be able to protect against the consequences of dehydration (Billi and Potts, 2002). In addition, survival of bacteria in the dry state is important in health issues (Berjak, 2006).

3.2 Mechanisms

Bewley (1979) pioneered the idea that desiccation-tolerance may be protoplasmic in a landmark article! The trait of desiccation-tolerance is an outcome of the interaction of a spectrum of phenomena and properties (Pammenter and Berjak, 1999) described as intrinsic cell characteristics (Walters *et al.*, 2005a). However, the degree of expression of these characteristics and, indeed, whether all or some are present underlie the differences in responses to drying between desiccation-tolerant and variously desiccation-sensitive organisms (Pammenter and Berjak, 1999). The modes and interaction of the operation of the protective mechanisms and processes of desiccation-tolerance remain largely conjectural (Berjak *et al.*, 2007; Berjak and Pammenter, 2008). Berjak (2006) stresses the point that desiccation-tolerance involves not only the facility to survive extreme water loss, but also the ability to survive for prolonged periods in the dehydrated state.

3.2.1 Intracellular physical characteristics

The first set of major components of the suite of mechanisms of desiccation-tolerance involves intracellular physical features (Pammenter and Berjak, 1999). It was first demonstarted by Berjak and co-workers (Berjak *et* al., 1984, 1989). They include minimization of vacuolation, protection of (the integrity of) DNA and orderly dismantling of cytoskeletal elements.

Both 'orthodox and recalcitrant seeds' deal with the problem of volume reduction by the accumulation of space-filling insoluble reserves (Berjak and Pammenter, 2008). There is the notable exception of 'highly-recalcitrant propagules' of *Avicennia marina*.

Boubriak *et al.* (2000) found DNA to be severely damaged after slight drying in embryos of *A. marina*. This event is followed by the inability for its repair after loss of 22 % of the water present at shedding. More recently, studies have shown putative role of helicases in plant abiotic stress tolerance (Owttrim, 2006; Vashisht and Tuteja, 2006).

With respect to the cytoskeleton, Faria *et al.* (2005) used an α-tubulin antibody in an immunocytochemical (ICC) assay to show that only disassociated tubulin granules were present in the radicle cells in the dry state of 'orthodox seeds' of *Medicago truncutula*. In contrast, well-established and extensive cortical arrays of microtubules were present in germinating *M. truncutula* and 'recalcitrant' *Inga vera* seeds (Faria *et al.*, 2005 and 2004, respectively). Microtubules disassociated on drying, giving rise to tubulin granules which disappeared following further dehydration. The damaged cells appeared to have lost the capacity for microtubule reconstitution upon rehydration. These observations support earlier findings indicating failure of the reconstitution of microfilaments and, hence, a complete cytoskeleton following desiccation in embryonic axes of *Quercus robur* (Mycock *et al.*, 2000) and *Trichilia dregeana* (Gumede *et al.*, 2003).

3.2.2 Intracellular de-differentiation and 'metabolic repression or switch-off or shut-down'

De-differentiation and 'metabolic switch-off or shut-down' constitute the second set of characteristics of the acquisition of desiccation-tolerance in 'orthodox seeds' (Pammenter and Berjak, 1999). Comparisons with developing 'recalcitrant seeds' indicate that these phenomena do not occur, although the metabolic rate may be at its lowest at, or shortly before, the seeds are shed (Farrant *et al.*, 1997).

Rogerson and Matthews (1977) observed a sharp decline in the levels of sugars, which preceded a fall in the respiratory rate, prior to the acquisition of desiccation-tolerance in developing seeds of *Pisum sativum*. Those authors suggested that such an event facilitated desiccation-tolerance in these tissues by, presumably, obviating metabolic damage.

Furthermore, Brunori (1967) showed that cell cycling was arrested at G1 phase during maturation drying in 'orthodox seeds'. The first round of S-phase replication occurred during G2 phase following imbibition. Desiccation-tolerance is lost as soon as cells enter G2M during which mitosis takes place (Sen and Osborne, 1974).

There is only a transient cessation of DNA replication at shedding, with re-entry into the S-phase soon thereafter, in embryos of *A. marina* (Boubriak *et al.*, 2000). In addition, the 4C DNA content was found to be relatively low and constant in both shoot and root apices in *I. vera* embryos from six weeks after flowering to shedding (Faria *et al.*, 2004). In contrast to the findings for *A. marina*, it did not change significantly after 13 h of imbibition of mature seeds. Reviewing previously published information for a range of 'recalcitrant-seeds species', Faria *et al.* (2004) concluded that the majority of cells appeared to be arrested in the G1 phase of the cell cycle, thus the more vulnerable 4C phase would be avoided when the seeds are shed and at the greatest risk of drying.

3.2.3 Sugars and oligosaccharides

It seems invariable that sucrose and certain raffinose series oligosaccharides or galactosyl cyclitols accumulate in 'orthodox seeds' during maturation drying as demonstrated by the pioneering studies of Leopold and co-workers and others (e. g. Blackman *et al.*, 1995; Steadman *et* al., 1996; Black *et al.*, 1999; reviewed by Leprince *et al.*, 1993; Horowicz and Obedendorf, 1994; Obendorf, 1997). In addition, high sucrose concentrations are also common to desiccated 'resurrection plant' tissues (reviewed by Berjak *et al.*, 1997).

Leopold and co-workers contend that the role of sucrose is dynamic in hindering the close approach of membranes to one another, and hence preventing their lateral proximity (Bryant *et al.*, 2001, Koster and Bryant, 2005; Halperin and Koster, 2006). In this regard, it is noteworthy that membrane lateral proximity promotes phase transitions of some phospholipids and even the demixing of membrane components which is accompanied by the exclusion of integral proteins (e. g. Ntuli *et al.*, 1997).

It appears that sucrose with raffinose or stachyose accumulates in the axes and cotyledons of developing 'recalcitrant seeds' (Berjak and Pammenter, 2008). 'Highly recalcitrant' embryos of *A. marina* were found to accumulate substantial amounts of sucrose and stachyose (Farrant *et al.*, 1993b). In addition, sucrose accumulation accompanied dehydration in the less 'recalcitrant' *Camellia sinensis* embryonic axes (Berjak *et al.*, 1989). Similarly, embryos of *Qurcus robur*, which are more desiccation-tolerant than those of the latter counterpart, accumulate sucrose and raffinose concomitant with the later stage of reserve accumulation (Finch-Savage *et al.*, 1993; Finch-Savage and Blake, 1994). Furthermore, *Quercus alba* embryos have a high sucrose content (Connor and Sowa, 2003).

From a wide-ranging survey of sucrose accumulation among both 'orthodox and recalcitrant seeds', it seems that a variety of 'recalcitrant seeds' accumulate substantial quantities of sucrose relative to oligosaccharide (Steadman *et al.*, 1996). However, Berjak and Pammenter (2008) argue that sucrose cannot play a part in protecting against desiccation damage as conjectured for 'orthodox seeds' as upon drying in the natural environment, 'recalcitrant seeds' would have already lost viability at WCs well in excess of those at which any benefits could be derived by the contribution of this disaccharide to the 'intracellular glass(y state)' (see below) or in counteracting lateral contact between membranes, as discussed above. It is probable that hydrolysis of sucrose affords a readily available respiratory substrate required to sustain ongoing development which grades imperceptibly into germination, followed by seedling establishment under favourable conditions in 'recalcitrant seeds' (Berjak and Pammenter, 2008).

3.2.4 Late embryogenesis-abundant/accumulating, (small) heat-shock proteins, oleosins and aquaporins

Galau and co-workers were the first advocates of the role of late embryogenesis-abundant/accumulating (LEA) proteins in desiccation-tolerance (Galau *et al.*, 1986, 1987; Galau and Hughes, 1987). LEA proteins, together with sucrose, have been the focus of much recent attention in the context of the acquisition and retention of desiccation-tolerance in 'orthodox seeds' (reviewed by Buitink *et al.*, 2002; Kermode and Finch-Savage, 2002; Berjak, 2006; Berjak *et al.*, 2007; Berjak and Pammenter, 2008).

Six groups of LEA proteins have been identified on the basis of particular peptide motifs (Cumming, 1999). These proteins generally lack cysteine residues, are composed predominantly of charged and uncharged polar amino acid and, with the exception of Group 5 LEA proteins, are highly hydrophilic and heat stable.

Buitink *et al.* (2006) have demonstrated that 18 genes coding for LEA and two heat-shock proteins (HSPs) were upregulated and identified as being common to the acquisition of desiccation-tolerance in *M. truncutula* seeds. The same situation prevailed during its experimental re-imposition in the seedlings.

It has been suggested that LEA proteins of some groups could provide a protective hydration shell around intracellular structures and macromolecules while others have been hypothesized to sequester ions during dehydration and in the desiccated state (Berjak and Pammenter, 2008). This action has been attributed to their hydrophilicity.

It has also been proposed that the lysine-rich K segment of Group 2 LEA proteins, dehydrins, might stabilize hydrophobic domains of other proteins which could become exposed as dehydration proceeds (Close, 1997). Such activity is ascribed to the propensity of dehydrins to form α-helices (Close, 1996). Such interactions could counteract inappropriate intermolecular hydrophobic associations (Cuming, 1999). A similar function has been suggested for small HSPs (sHSPs) (reveiwed by Buitink *et al.*, 2002; Berjak *et al.*, 2007).

What is especially significant in terms of desiccation-tolerance is that dehydration, particularly in the presence of sucrose, induces at least some LEA proteins to assume α-helical conformation (Wolkers *et al.*, 2001). Such conformational change is suggested to be the basis of the formation and maintenance of the intracellular 'glass(y state)' (Berjak, 2006, Berjak *et al.*, 2007, Berjak and Pammenter, 2008). An additional feature linking LEA proteins to desiccation-tolerance is their concomitant appearance with abscisic acid (ABA) regulation of *lea* gene transcription (reviewed by Bray, 1993; Kermode, 1990, 1995; Cuming, 1999; Kermode and Finch-Savage, 2002; Berjak, 2007).

Berjak and Pammenter contend that although the evidence for LEA protein involvement in desiccation-tolerance is 'correlative and circumstantial rather than by direct experimental demonstration' (Cumming, 1999), it is compelling and, indeed, convincing! The appearance of LEA proteins is associated with 'orthodox seed' maturation, as it is with the imposition of a variety of stresses causing water deficits in plant cells (Cumming, 1999).

The situation regarding the occurrence of LEA proteins in 'recalcitrant seeds' is equivocal (Berjak and Pammenter, 2008). They occur in a range of species from different habitats while apparently being absent from others. Group 2 LEA proteins, dehydrins, have been identified in 'recalcitrant seeds' of some temperate trees (Finch-Savage *et al.*, 1994; Gee *et al.*, 1994), other temperate species and some of (sub-)tropical provenance (Farrant *et al.*, 1996) and in grasses typified by *Portersia coarctata*, *Zizania* spp. and *Spartina anglica* (Gee *et al.*, 1994). However, no dehydrin-type LEA proteins could be found in seeds of ten tropical wetland species (Farrant *et al.*, 1996).

Berjak and Pammenter (2008) argue that it is difficult to envisage a functional role for LEA proteins in 'recalcitrant seeds' based on the conjecture about functionality of such proteins in desiccated or desiccating 'orthodox seeds'. However, those authors maintain that the presence of LEA proteins in 'recalcitrant seeds' of particular species could facilitate more effective survival to lowered water contents following extremely rapid dehydration by flash drying necessary to enable the axes to be cryopreserved (Berjak *et al.*, 1990).

Collada *et al.* (1997) showed the abundant presence of small HSPs (sHSPs) in cotyledons of 'recalcitrant' *Castenea sativa*. Those authors thus concluded that the occurrence of sHSPs could not be linked to desiccation-tolerance. However, a variety of unidentified sHSPs are expressed in 'recalcitrant' amaryllid embryos, most of which are amenable to cryostorage (Berjak and Pammenter, 2008).

Leprince and co-workers were the first to implicate oleosins in desiccation-tolerance (Leprince *et al.*, 1998). Oleosins in hydrated cells are held to maintain oil bodiers as discrete entities. They were suggested to be lacking in or inadequate in lipid-rich 'recalcitrant seeds'. In this regard, no oleosins were detected in the 'highly recalcitrant seeds' of the tropical species, *Theobroma cacao*. However, later work involving cloning and characterization of cDNA and peptide sequencing has shown that two oleosins are present in mature *T. cacao* seeds (Guilloteau *et al.*, 2003).

Major intrinsic proteins (MIPs) are a family of channel proteins that are mainly represented by aquaporins (APs) in plants. They are generally divided into tonoplast intrinsic proteins (TIPs) and plasmalemma intrinsic proteins (PIPs) according to their subcellular localisation (reviewed by Maurel *et al.*, 1997).

For instance, the vacuolar membrane protein, α-TIP, a water channel protein accumulates during seed maturation in parenchyma cells of seed storage organs. Synthesis of this integral membrane protein does not appear to be related, in a quantitative manner, to storage protein deposition. A role in seed desiccation, cytoplasmic osmoregulation and/or seed rehydration has been suggested (Johnson *et al.*, 1989).

The water-channel activity can be regulated by phosphorylation. The protein assembly as a 60 Å X 60 Å square in which each subunit is formed by a heart-shaped ring comprised of a-helices (Daniels *et al.*, 1999). Homologues to PIPs and TIPs are controlled by dehydration and ABA in desiccation-tolerant resurrection plant *Craterostigma plantagineum* (Mariaux *et al.*, 1998). Members of a subset of PIPs (PIPa) are regulated by ABA-dependent and ABA-independent pathways.

3.2.5 Intracellular 'glass(y [/vitrified] state)'

There is considerable evidence for the existence of the intracellular milieu in the 'glassy or vitrified state' in 'orthodox seeds' once sufficient water has been lost (Berjak and Pammenter, 2008). Leopold and co-workers pioneered the involvement of (the) 'glass(y state), as a consequence of a supersaturated sugar solution, in desiccation-tolerance (Koster and Leopold, 1988; Williams and Leopold, 1989).

Later, there was a realization that there are many other intracellular molecules that must contribute to 'glass' (e. g. Walters, 1998). Koster (1991) was the first to show that certain properties of model systems, constituted to simulate intracellular sugar mixtures, differed from the situation in seeds.

In addition, Oliver *et al.* (2001) suggested that LEA proteins might underlie the stability of intracellular 'glasses' in the dry state. Existing as unordered random coils in solution, LEA proteins assume a far more ordered conformation upon dehydration as demonstrated, for example, by Wolkers *et al.* (2001) for a Group 3 LEA protein from desiccation-tolerant *Typha* pollen and Boudet *et al.* (2006) for both Group 1 and Group 5 LEA proteins from *M. truncutula* seeds.

Berjak (2006) proposed that intracellular 'glasses' in dry seeds may be based on coiled LEA proteins in interaction with sucrose and the residual water. However, a sugar-based phase might occur in narrow intermembrane spaces, with LEA proteins excluded on the basis of size as suggested by Koster and Bryant (2005).

While the relative stability of intracellular 'glassy state' is held to maintain viability albeit not indefinitely in the desiccated state of 'orthodox seeds', intracellular 'glasses' just would not normally form in 'recalcitrant seeds' as they require water contents of less than 0.3 g g^{-1} (Berjak and Pammenter, 2008). Water concentrations of approximately 0.3 g g^{-1} coincide with a marked increase in cytomatrical or cytoplamic viscosity, indicative of 'glass formation' (Buitink and Leprince, 2004).

Under slow drying conditions which would prevail in the natural environment, 'recalcitrant seeds' die at far higher water contents (reviewed by Pammenter and Berjak, 1999; Walters *et al.*, 2002)! However, it may be possible that transient intracellular 'glasses' can be formed as a consequence of flash drying of excised embryonic axes, which is a procedure intrinsic to the cryopreservation protocol for germplasm conservation of 'recalcitrant-seeded' species (Berjak and Pammenter, 2008).

3.2.6 (Re)active oxygen species and free radical-processing antioxidants

The free-radical theory of ageing originated in the medical sciences more than half-a-century ago (Harman, 1956). It was later introduced into seed science when Kaloyereas (1958) suggested that lipid oxidation might underlie loss of viability in seeds.

There has been a particular focus on free radicals, (re)active oxygen species (R/AOS) and antioxidant systems implicated in the acquisition and maintenance of desiccation-tolerance in both 'orthodox seeds' and vegetative tissues of 'resurrection plants' of late (Berjak and Pammenter, 2008). One of the most intriguing aspects of ROS to have emerged recently is their dual role in intracellular signaling as well as intracellular destruction (reviewed by Laloi *et al.*, 2004; Foyer and Noctor, 2005; Suzuki and Mittler, 2006).

Free radicals and AOS are held to result from metabolic imbalance in cellular respiration and photosynthesis. For example, phosphofructokinase (PFK), a rate-limiting enzyme in glycolysis, malate dehydrogenase (MDH), a key enzyme in the tricarboxylic acid (TCA) cycle and dehydrogenases of complexes I and IV of the electron transport chain all of oxidative phosphorylation are slightly, midly and highly adversely affected by desiccation (reviewed by Côme and Corbineaeu, 1996).

AOS are formed when high-energy electrons are transferred to molecular oxygen (O_2). They include singlet oxygen ($^1O_2$), hydrogen peroxide (H_2O_2) and the superoxide (O_2^-) and the hydroxyl ($OH\cdot$) radicals. They have long been considered toxic species that can cause damage to lipids, protein and nucleic acids (e. g. Halliwell, 1987; Fridovich, 1998; Hendry, 1993; Suzuki and Mittler, 2006). Not surprisingly, the activity of a spectrum of enzymatic and non-enzymatic antioxidants is considered to be of prime importance in quenching ROS activity.

However, ROS are now considered as secondary messengers in a diversity of signal transduction cascades in metabolically active hydrated plant tissues (Foyer and Noctor

2005). Hydrogen peroxide and the superoxide radical are singled out because of their implication in many plant developmental and growth processes. Nevertheless this role does not gainsay the vital necessity of their control by a spectrum of antioxidants (Berjak and Pammenter, 2008).

While strict control of AOS is taken for granted in hydrated cells, possession and effective operation of a suite of both enzymatic and non-enzymatic antioxidants is of prime importance during dehydration of 'orthodox seeds' and desiccation-tolerant vegetative tissues (e. g. Kranner *et al.*, 2002; reviewed by Pammenter and Berjak, 1999; Bailly, 2004; Kranner and Birtić, 2005; Berjak, 2006; Berjak *et al.*, 2007; Berjak and Pammenter, 2008). This scenario prevails in the dry state and as soon as water uptake by desiccated cells commences.

Antioxidants are either enzymic or non-enzymic. Enzymic oxidants include ascorbate free radical reductase (AFRR), ascorbate and guaicol peroxidase (A/GPO[D]), catalse (CAT), dehydroascorbate reductase (DHAR), glutathione reductase (GR) and superoxide dismutase (SOD). Non-enzmic oxidants consists of ascorbate/ic acid (AsA) (vitamin C), reduced and di-/oxidized glutathione (GSH and GSSG), retinol (vitamin A) and α, β and γ-tocopherol (vitamin E).

It is possible that certain antioxidants may be operative within localized regions of higher water activity within desiccated cells. As an example, 1-cys-peroxiredoxin (CPR) has been localized to nuclei in imbibed dormant barley embryos (Stacey *et al.*, 1999). It has been suggested to provide antioxidant protection to DNA.

However, it has been suggested that there are localized regions with water activity adequate to facilitate molecular mobility in the desiccated state – 'localised water pools' (Rinne *et al.*, 1999; Leubner-Metzger, 2005). Berjak (2006) argues that it is possible that CPR can function to protect the genome against ROS in desiccated seeds if some such regions occur in the mileu of the chromatin. In this respect, it should be remembered that the cysteinyl residue of CPR can be regenerated ultimately by electron donors such as thioredoxins and glutaredoxin (Dietz, 2003).

In this regard, Rinne *et al.* (1999) conjectured that enzyme activity continues to occur in dehydrin-associated areas of greater water activity in the otherwise dehydrated cells of buds. Leubner-Metzger (2005) showed localized β-1,3-glucanase activity in the inner testa to be instrumental in after-ripening in air-dry tobacco seeds. A similar argument may be advanced for the activity of other enzymic, as well as non-enzymic, antioxidants (Bailly, 2004) in localized regions of greater water activity within intracellular 'glasses' in dehydrated seeds (see above).

Unlike the situiation in 'orthodox seeds' during the latter stages stages of development, metabolism is sustained at measurable levels in 'recalcitrant seeds' (Farrant et al., 1993a, b). When water is lost, and especially when dehydration proceeds slowly, metabolism is considered unbalanced. This situation can result in considerable intracellular metabolic damage and death of seed/embryos at surprisingly high WCs (Pammenter and Berjak, 1999 and Walters *et al.*, 2002; Ntuli, 2011a) (see below). Metabolic damage in 'recalcitrant seeds' is thought to be intimately associated with the generation of AOS where the intracellular antioxidant defences are inadequate to quench them.

Recent data for 'recalcitrant' *Araucaria bidwilli*, *Quecus robur*, *Trichilia dregeana* and germinating *Pisum sativum* embryos show that there is a transient increase in antioxidant activity upon initial dehydration. However, with further water loss, activity declines, accompanied by an increase in free radicals and thio-barbituric acid-reactive substances (TBARS) (Francini *et al.* 2006; Ntuli *et al.*, 2011; Song et al., 2004; Ntuli, 2011b, respectively).

3.2.7 Osmolytes/protectants

It is now established that compatible organic solutes play a central role in plant drought tolerance (Ashraf and Foolad, 2007). Overproduction of compatible organic osmotic is one of the responses of plants exposed to osmotic stress (Serraj and Sinclair, 2002; Ashraf *et al.*, 2008).

Among the many organic osmolytes known to play a substantial role in stress tolerance, glycine betaine (GB), a quaternary ammonium compound, occurs richly in response to dehydration stress (Mansour, 2000; Mohanty *et al.*, 2002; Yang *et al.*, 2003; Ashraf and Foolad, 2007). Choline monooxygenase (CMO) and betaine aldehyde dehydrogenase (BDH) are two key enzymes for the biosynthesis of GB in higher plants.

Proline, like GB, is also an important compatible organic osmolyte that plays a key role in stress tolerance (reviewed by Ashraf, 2010). Pyrroline-5-carboxylate synthetase (PCS) is the key enzyme for proline biosynthesis.

Trehalose, a nonreducing sugar, is also a potential organic osmoticum which has a substantial role in the protection of plants against stresses (reviewed by Ashraf, 2010). Trehalose-6-phosphate synthase (TPS) is a key enzyme involved in trehalose biosynthesis.

Mannitol, a polyol, is one of the most important osmoprotectants that play a vital role in plant stress tolerance (reviewed by Ashraf, 2010). Mannitol-1-phosphate dehydrogenase (MPD) is involved in mannitol biosynthersis.

3.2.8 Lipid composition

Liu *et al.* (2006) showed that the proportion of saturated fatty acids in membrane phospholipids was significantly higher in 'recalcitrant' than in 'orthodox seeds'! In addition, Nkang *et al.* (2003) found that mature seeds of *Telfairia occidentalis*, which were characterized by predominantly saturated fatty acids, increased accumulation of both mono- and polyunsaturated fatty acids when dried at 28 °C when total lipid was evaluated. In contrast, high levels of saturated fatty acids were retained and the marked decline in viability was delayed when seeds were dried at 5 °C. Interestingly, Ajayi *et al.* (2006) reported that *T. occidentalis* seeds lost viability at 6 °C within a relatively short time, suggesting chilling sensitivity, despite the retention of saturated fatty acids during desiccation at 5 °C reported by Nkang *et al.* (2003).

Lipid peroxidation has been shown to be associated with deterioration of seeds exhibiting 'intermediate' post-harvest physiology, particularly in terms of chilling sensitivity (Berjak and Pammenter, 2008). Crane *et al.* (2006) have shown that crystallization of the predominantly saturated storage lipid occurs in *Cuphea carthagenensis* seeds at both high and very low WCs after maintenance at 5 °C. Those authors showed that rehydration without a preceding melting of crystallized triglycerides was lethal. Similarly, Neya *et al.* (2004)

showed that hydrating lipid-rich 'non-orthodox seeds' of *Azadirachta indica* in warm water alleviated the effects of imbibitional stress that occurred when cold water was used.

3.2.9 Repair

Osborne and co-workers were the first to advocate the role of DNA repair in desiccation-tolerance (e. g. Osborne, 1983). 'Orthodox seeds' must apparently repair any damage accumulated in the dry state soon after imbibition is initiated. Repair occurs in the lag phase of water uptake before radical protrusion.

This requirement demands unimpaired operation of repair mechanisms and restitution of normal cell structure and function. Simultaneoulsy, the presence and efficient operation of appropriate antioxidants is vital (reviewed by Pammenter and Berjak, 1999).

Newly harvested *A. marina* embryos were capable of repair when DNA fragmentation was induced by radiation but this ability was increasingly compromised if embryos had first been dehydrated (Boubriak *et al.*, 2000). In addition, Connor and Sowa (2003) showed that a reversible shift occurred between the gel and liquid crystalline phases on rehydration after initial dehydration of *Q. alba* acorns form Fourier transform infra-red analyses of membrane lipids.

3.2.10 Endogenous amphiphilic substances

Hoekstra and co-workers pioneered the implication of amphiphilic substances in desiccation-tolerance (Golovina *et al.*, 1998). Those authors showed that dehydration of tolerant pollen and seeds has the potential to cause certain amphiphilic molecules to migrate into membranes, with migration back into the cytomatrix upon rehydration. The amphiphiles were suggested to play a role in maintaining core fluidity of membranes in the dry state. However, subsequent investigations could not confirm this phenomenon. Although the amphiphiles may fluidise membranes, correlation with desiccation-tolerance was uncertain (Goloviva and Hoekstra, 2002).

3.2.11 Control of acquisition and maintenance of desiccation-tolerance

The phenomenon of the control of the acquisition and maintenance of desiccation-tolerance in 'orthodox seeds' is suggested to be pre-programmed and developmentally regulated and initiated by maternal factors rather than directly via environmental signals and later to be under the control of the embryo (reviewed by Bewley and Black, 1994; Bewley, 1979, 1997). A major point of confusion is that desiccation-tolerance overlaps with other maturation processes and the development of dormancy where it occurs.

Studies on viviparous mutants and those characterized by impairment of the maturation process in *Zea mays* (*VP* series mutants) and *Arabidopsis thaliana* (*LEC1, LEC2, FUS3* and *ABI3*) have indicated both ABA-independent and ABA-dependent pathways of gene regulation (e. g. Bray, 1993; Kermode, 1990, 1995; Kermode and Finch-Savage, 2002; Bartels, 2005; Vincente-Carbajosa and Carbonero, 2005).

Current understanding of the control of seed maturation and acquisition of desiccation-tolerance in *A. thaliana* suggests that *LEC1, LEC2, FUS3* and *ABI3* are the four 'master genes' involved, with the latter three implicated in desiccation-tolerance (To *et al.*, 2006). Another pivotal factor is ABA which, in the context of seed development, is probably best known for its role in regulating *lea* gene transcription (e. g. Bray, 1993; Kermode, 1990, 1995; Cuming, 1999; Kermode and Finch-Savage, 2002).

Furthermore, recent evidence suggests that a delicate balance between various ROS, as secondary messengers, and antioxidants may be intimately involved with seed maturation and the acquisition of desiccation-tolerance (Berjak *et al.*, 2007). In addition, there is interaction and cross-talk in the operation of the factors involved!

4. Desiccation-sensitivity

Besides producing short-lived seeds, many of the 'recalcitrant-seeded species' are threatened by overexploitation, indiscriminate harvesting and habitat loss (Berjak, 2005). Hence, understanding the phenomenon of 'seed recalcitrance', and consequently developing sound conservation practices for species producing such seeds, is of major scientific and practical importance (Berjak and Pammenter, 2008). Such importance is underscored in Target VIII of the Global Strategy for Plant Conservation of the Convention on Biodiversity which calls for 60% of all threatened species to be in accessible *ex situ* collections by 2010 (Berjak and Pammenter, 2008)!

'Recalcitrant seeds' remain sensitive to drying both during development and after they are shed from the parent plant. However, the range of WCs of the embryonic axes when seeds are shed varies markedly among species – 0.4 to 4.4 g g^{-1} (Chin and Roberts, 1980 ; Berjak and Pammenter, 2004). Some decline in WC prior to shedding has been recorded for several temperate species (e. g. *Acer pseudoplatanus* [Hong and Ellis, 1990], *Aesculus hippocastum* [Tompsett and Pritchard, 1993] and *Quercus robur* [Finch-savage and Blake, 1994]) and some (sub-)tropical provenance (e. g. *Machilust hunbergii* [Lin and Chen, 1995]) and species (e. g. *Ekerbergia capensis* [Berjak and Pammenter, 2008]), leading to the suggestion that a measure of desiccation-tolerance might be acquired during development (Finch-Savage and Blake, 1994), with no further importation of water (Berjak and Pammenter, 2000).

Nevertheless, even for those seeds that are shed at axis WCs towards the lower end of the range, further dehydration is deleterious, indicating that at least some of the mechanisms necessary for complete desiccation-tolerance are not entrained (Berjak and Pammenter, 2008). In contrast, WCs of axes of recalcitrant seeds of most of (sub-)tropical species which have been investigated lie at the high end of the range (> 1.5 g g-1), and the axes are damaged after only slight drying, particularly if water loss is slow (Berjak and Pammenter, 2008). Those authors argue that this observation indicates that few, if any, of the mechanisms putatively affording 'orthodox seeds' tolerance to desiccation are operational.

The degree of 'recalcitrance' may be difficult to quantify (Pammenter *et al.*, 2002a). Nonetheless, seeds of *Avicennia marina* (Farrant *et al.*, 1993a, b) and *Hopea* species (e. g. Chin and Roberts, 1980 ; Sunilkumar and Sudhakara, 1998) are considered highly 'recalcitrant'. In this regard, studies on *Zizania* spp. indicated that, despite differences in WCs at which desiccation damage occurred in embryos at different developmental status, all equated to a common water activity value of 0.90 (Vertucci and Farrant, 1995) ! These observations may be related to a common spectrum of metabolic events that are impaired, which has been suggested to occur in particular water potential ranges, as 'recalcitrant seeds' are dried (Vertucci and Farrant, 1995).

Until relatively recently, seed screening for non-orthodox behaviour was based initially on water content of ostensibly mature seeds at shedding, followed by ascertaining viability retention upon sequential removal of increasing proportions of tissue water (Hong and Ellis, 1996 ; International Plant Genetic Resources Insitute [IPGRI]/DFSC, 2004. However, because frequently only small numbers of recalcitrant seeds are able to be harvested at any one time, Pritchard *et al.* (2004a) developed the 100-seed test which proved to be a reliable indicator of the desiccation responses of seeds of eight tropical palm species. Berjak and Pammenter (2008), nonetheless, recommend modofcations of the IPGRI/DFSC protocol (2004) whenever seed numbers allow. The IPGRI/DFSC protocol includes an assessment of viability retention in storage at a range of temperatues. This approach is time-consuming, but give a reliable indication of whether seeds are 'orthodox', 'recalcitrant' or fall somewhere between 'orthodoxy' and 'recalcitrance'.

Nevertheless, a variety of studies (reviewed by Daws *et al.*, 2006b) have indicated that there could be a correlation among 'recalcitrance', seed characteristics and variables characterising individual habitats. Acting on these indications, and analysing several of the parameters across 104 tropical species from 37 families deriving from Panama, Daws *et al.* (2006b) developed a relaibly predicitve model based on the just two of the traits: seed mass and seed coat/coverings ratio ([SCR] i. e. the ratio of seed coverings mass to mass of the whole seed). Desiccation-sensitivity was found to be significantly related to relatively low SRCs, typified by large seed size coupled with thin coverings. The predictive value of the model was convincingly shown when it was further applied to seeds of 28 African species and ten species from Europe, showing in all cases that the predicition was in agreement with published data on responses of the seeds to drying (Berjak and Pammenter, 2008).

The original categorisation of seeds according to their post-harvest storage responses, which embodies the idea of two distinctive groupings – 'orthodox' and 'recalcitrant', was pioneered by Roberts (1973) in a landmark paper! A further category – intermediate – was later introduced (Ellis *et al.*, 1990a). Intermediate storage behaviour implies that the seeds are shed at relatively high WCs, but will withstand considerable dehydration, although not to the extent tolerated by 'orthodox seeds', especially at low temperatures.

Although categorising seeds into the three distinct categories is useful, it is noteworthy that desiccation-sensitivity can be further subcategorised into three subcategories – highly, moderately and minimally – 'recalictrant' (Farrant *et al.*, 1988). Berjak and co-workers and many now favour an open-ended continuum of seed behaviour, subtended by extreme 'orthodoxy' at the one end and the highest degree of 'recalcitrance' at the other (Berjak and Pammenter, 1994, 1997a, b, 2000, 2001, 2004, 2008; Pammenter and Berjak, 1999 ; Sun, 1999 ; Kermode and Finch-Savage, 2002, Berjak, 2005, 2006).

The continuum concept accomodates the marked variability occuring both between and within species and is supported by evidence indicating that provenance has a significant effect on seed development and the degree of drying tolerated for individual species (Daws *et al.*, 2004a,b 2006a). In terms of the continuum concept, it is considered pertinent that even 'orthodox seeds' are not equally desiccation-tolerant!

4.1 Occurrence

Chin and Roberts (1980) published the first list of species recorded as producing 'recalcitrant seeds', and collated what was then known about their post-harvest behaviour. The species those authors listed produce seeds important in agroforestry (e. g. species of *Quercus* and *Shorea*), as crops for seed or fruit consumption (e. g. *Castanea* species and *Artocarpus heterophyllus*), or commodity production (e. g. *Elais guineensis, Havea brasiliensis* and *Theobroma cacao*).

Since then, the seed biology of a range of lesser known and generally (sub-)tropical tree species has been studied, revealing many more produce 'recalcitrant or otherwise non-orthodox seeds' (reviewed by Berjak and Pammenter, 2004; Sacandé *et al.*, 2004; Flynn *et al.*, 2006). In a single year, a screening programme focusing on southern African species revealed that seeds of at least 17 herbaceous geophytic amaryllids are 'recalcitrant' – which is unusual in indicating a familial trait – while seeds of a further 13 unrelated trees species have also proved 'recalcitrant' (Erdey *et al.*, 2007)!

While the majority of the species producing 'recalcitrant seeds' are endemic to the humid (sub-)tropics, such seeds are also produced by a small spectrum of mainly trees of temperate provenance, while certain dryland species (Danthu *et al.*, 2000 ; Gaméné *et al.*, 2004 ; Pritchard *et al.*, 2004b) have also been identified as being 'recalcitrant seeded'. Studies have also identified seeds of some cycads (e. g. *Encephalartos* spp.) to be 'recalcitrant' (Woodenberg *et al.*, 2007), while Daws *et al.* (2007) reported a considerable incidence of desiccation-sensitive seeds among palms.

4.2 Variability

There are marked differences in the degree of drying that 'recalcitrant seeds' will tolerate, although the lowest WC survived depends on other parameteres, especially the rate at which water is lost (see below) (reviewed by Berjak and Pammenter, 2008). Comparisons of published data on individual species are often not helpful, because of the differing conditions under which dehydration was carried out.

However, a similar pattern of ultrastructural events terminating in cell lysis was recorded as occuring at markedly different WCs when 'recalcitrant seeds' of three unrelated species – a gymnosperm: *Araucaria angustifolia*, a dicotyledonous vine: *Landolphia kirkii* and a herbaceous monocot: *Scadoxus membranaceous* - were dried under identical conditions (Farrant *et al.*, 1989). Significantly, the rate at which the seeds of the three species lost water was inversely related to the WC at which viability was lost!

4.2.1 Genera

Differences in the crtical/lethal/lowest 'safe' WC which 'recalcitrant seeds' will withstand are not confined to disparate genera (reviewed by Berjak and Pammenter, 2008). They have also been noted for different species of individual genera.

A thought-provoking finding is that seeds of different species of a single genus may be differently categorised (reviewed by Berjak and Pammenter, 2008). For example, species of *Acer* and *Coffea* are variously categorised (Hong and Ellis, 1990 ; Eira *et al.*, 1999).

4.2.2 Provenance

Interestingly, recent data indicate that seeds of *A. hippocastum* and *A. pseudoplatanus* from different provenances differ in their response to dehydration (Daws *et al.*, 2004b ; 2006a, respectively).

4.2.3 Developmental status

While developing seeds of *Machilus thunbergii* lost viability within thirty days when dried at 73% RH and 25 °C, those that were mature were able to tolerate a 19% loss of water before germinability declined (Li and Chen, 1995). Differing degrees of desiccation sensitivity have been similarly correlated with developmental status for *Landolphia kirkii* and *Camellia sinensis* (Pammenter *et al.*, 1991; Berjak *et al.*, 1992; 1993).

It appears that the least desiccation-sensitive stage generally occurs when the metabolic rate is the lowest which usually but not invariably coincides with shedding (reviewed by Berjak and Pammenter, 2008). However, desiccation sensitivity increaeses markedly as germinative metabolism progresses (Farrant *et al.*, 1986 ; Berjak *et al.*, 1989).

4.2.4 Tissues

Zygotic embryonic axes and storage tissues seldom have if ever have the same WC (e. g. *Acer hippocstum* [Tompsett and Pritchard, 1993]). There is also an uneven water distribution between component tissues in *Araucaria hunsteinii* (Pritchard *et al.*, 1995).

Frequently, axes are at higher WCs, and more desiccation-sensitive than cotyledons (reviewed by Berjak and Pammenter, 2008). However, cotyledons have been reported to be more sensitive to dehydration than axes in *Castanea sativa* (Leprince *et al.*, 1999).

4.2.5 Season

A further contribution to the variability among seeds of individual species is that their characterisitcs differ both intra- and interseasonally (reviewed by Berjak and Pammenter, 2008). Intraseasonal variation includes differing WCs of the component tissues of ostensibly mature seeds depending on the time of harvest (reviewed by Berjak and Pammenter, 1997a,b).

An additional feature that has been consistently been observed is the poor quality of seeds produced late in the season, which are often severely fungally infected. In this regards, an enhanced rate of deterioration upon dehydration has been reported for late-harvested seeds of *Machilus kanoi* (Chien and Lin, 1997). It has also been observed that late-season fruits of *Avicennia marina* and *Syzygium cordatum* have a tendency either to abort or not abscise.

It is probable that the poor quality of late-season seeds may be explained in terms of the cumulative heat sum during development (Daws *et al.*, 2004b). Those authors monitored *Acer hippocastum* seed development along a latitudinal gradient and reported that the greater the cumulative heat sum, the more robust, further developed and less desiccation sensitive were the seeds.

Berjak and Pammenter (2008) argue that a similar interpretation for poor seed quality can be applied to fruits and seeds produced in the latter part of the season in non-equitorial zones.

Temperatures decline as the summer wanes and, accompanied by by shortening day-lengths, results in a sob-optimal heat sum to late-developing fruits. This phenomenon is proposed to influence fruit and seed development negatively, resulting in their poor quality, which includes lowered resistance to fungal establishment.

Interseasonal varation among seeds of the same spcies may be similary rationalised (Berjak and Pammenter, 2008). For example, *Camellia sinensis* seeds harvested showed axial WCs as disparate as 2.0 ± 0.3 to 4.4 ± 2.4 g g^{-1} (Berjak *et al.*, 1996).

4.2.6 Individual seeds

There are usually marked differences in axial WCs among individual seeds (reviewed by Berjak and Pammenter, 1997a,b)! These differences persist even when seeds are harvested simultaneously.

4.3 Experimental conditions

A number of experimental parameters determine survival during drying, cooling, warming or thawing and imbibition or rehydration of seed tissues.They include drying rate and temperature, cooling rate and imbibition or rehydration rate and temperature.

Disparate opinions have been expressed about the effect of drying rate on the critical/lethal/lowest WC tolerated by 'recelacitrant seeds without compromising viability (reviewed by Berjak and Pammenter, 2008). However, it is apparent that the actual rates described as raspid or slow can pertain to very different time scales.

For example, rapid desiccation can be achieved in a matter of as little as fifteen minutes for some species and as much as twent-four hours for others when dehydrating excised axes by 'flash drying' (e. g. Ntuli and Pammenter, 2009). Both these examples are rapid relative to axes within whole seeds which require a matter of days to attain similarly low WCs (e. g. Pammenter *et al.*, 1998).

Zygotic embryonic axes generally but not invariably consitute a very small proportion of the total mass or volume of a 'recalcitrant seed'. When excised, they can be dried very rapidly in a laminar air-flow or using the technique of 'flash-drying' (reviewed by Pammenter et al, 2002b; Ntuli, 2011c).

Rapidly-dried axes and, occassionally, seeds will survive to far lower WCs than can be attained on slow dehydration (reviewed by Berjak and Pammenter, 2008). This phenomenon has been shown for a variey of species. Fast desiccation facilitates axis viability retention well into hydration level III and, occassionally, just into level II (Vertucci and Farrant, 1995), at which extreme, generally lethal damage is associated with slow water loss.

It is not that 'flash-drying' renders 'recalcitrant seeds' desiccation-tolerant (reviewed by Berjak and Pammenter, 2008). On the contrary, they will rapidly lose viability at ambient or refridgerator temperatures if allowed to remain at the low WCs attained. 'Flash-drying' achieves the rapid passage through the intermediate WC ranges at which aqueous-based metabolism-linked - metabolic damage - occurs.

'Recalcitrant seeds' are metabollically active and initiate germination around shedding (reviewed by Berjak and Pammenter, 2008). As a result, their developmental status is becoming more advanced and their desiccation sensitivity is increasing. Thus, desiccation-sensitivity could be increasing as drying is proceeding if germination is occurring at the same time frame as drying so reducing the water loss tolerated (Berjak et al., 1984, 1989; Farrant et al., 1986).

The drying rate is markedly affected by the nature of the coverings, size and developmental status (reviewed by Berjak and Pammenter, 2008). There is a lower WC limit below which 'recalcitrant seeds' will not survive. This WC is generally at or near the level at which all the remaining water is structure associated. Desiccation damage sensu stricto ensues when structure-associated or non-freezable water. A major differnce between orhtodx and recalcitrant seeds' is that the former can lose a considerable proportion of this water.

4.4 Storage

Seed storage is imperative, not only to provide good-quality planting material and feedstock from season to season in agriculture, as well as interseasonal food reserves, for food security, but also as base and active collections in the long term conserving of genetic resourses for biodiversity conservation (reviewed by Berjak and Pammenter, 2008). As long as 'orthodox seeds' are of high quality after harvest, the period for which they can be stored without deterioration is predictable under defined conditions of low temperature and RH that will maintain low WC.

Storage longevity of 'orthodox seeds' increases logarithmically with decreasing WC (Ellis and Roberts, 1980) although there appear to be limits of drying beyond which no further advantage is gained (Ellis et al., 1990b) and, in fact, if exceeded, may be damaging (e.g. Walters, 1998 ; Walters and Engels, 1998 ; reviewed by Berjak, 2006). However, Berjak and Pammenter (2008) noted that there is no unanimity about this issue (Hong et al., 2005). Nonetheless, 'orthodox seeds' have finite lifespans – years, decades or centuries – even under ideal conditions (Walters et al., 2005b).

In contrast, 'recalcitrant seeds' are characterised by post-harvest lifespans of the order of days to months, or for temperate species, a year or two, as long as such seeds will tolerate low but not sub-zero temperatures (e. g. Chin and Roberts, 1980). The inexoreable progress of germinative metabolism – which occurs with no requirement for additional water – is one of the major factors hampering short- to medium-term storage of 'recalcitrant seeds'. The developmental status of 'recalcitrant seeds' changes rapidly after they are shed because they not only hydrated but metabolically active.

Enzymic antioxidants have been found to be inadequate in counteracting oxidative stress during storage (Tommasi et al., 2006) in 'recalcitrant seeds' of Ginkgo biloba (Liang and Sun, 2002). The viability of the seeds at 25 °C declined from 80 to 46% between three and six months, accompanied by a decline in WC from c. 2.0 to 1.0 g g^{-1}. The seeds stored at 4 °C lost viability precipitously between six and nine months in storage during which the reduction in WC was insignificant.

It is suggested that the viability at 25°C was the direct result of metabolic damage (Pammenter and Berjak, 1999; Walters et al., 2001, 2002) (see above). Generation of free radicals/ROS and accumulation of thiobarbituric acid-reactive substances (TBARS), along with the decreasing ability of antioxidants to modulate the situation, is consistent with the water stress-induced damage in both cases. It is proposed that death occurred at 4 °C because metabolism progressed albeit slowly to the stage at which exogenous water supply was needed (e. g. Berjak *et al.*, 1989; Pammenter *et al.* 1994; Ntuli, 2011d). Working with 'recalcitrant' *Acer saccharinum* seeds, Ratajczak and Pukacka (2006) concluded from changes in enzymes of the ascorbate-glutathione cycle and levels of ascorbate and glutathione that viability of stored hydrated seeds could be maintained only when a vigorous antioxidant system was operational.

4.4.1 Wet storage

The only way in which vigour and viability of 'recalcitrant seeds' can be maintained is to keep them under conditions not permitting water loss at the lowest temperature they will withstand and to eliminate or, at least, minimise seed-associated mycoflora (reviewed by Berjak and Pammenter, 2008). Nevertherless, storage of whole seeds is strictly short- to medium-term. This limitation is due to the fact that 'recalcitrant seeds' are metabolically active, and will progress from development into germination at shedding.

'Recalcitrant seeds' of some species are shed consireably before development is complete (reviewed by Berjak and Pammenter, 2008). For example, *Trichilia dregeana* seeds can be stored for several months at 16 °C (Goveia *et al.*, 2004).

Storage longevity may be further optimised in the case of 'recalcitrant seeds' that are not chilling sensitive which would be expected for temperate species (reviewed by Berjak and Pammenter, 2008). Chilling sensitivity may be provenance related, and there appears to bedistinct genetic differences among plants from different provenances (Bharuth *et al.*, 2007).

It has been sporadically suggested that lowering 'recalcitrant seed' WC to levels permitting basal metabolism but precluding germination in storage might extend their longevity (reviewed by Berjak and Pammenter, 2008). This means has proved to be deleterious to both life span and quality of seeds for a range of species (Corbineau and Côme, 1986a, b, 1988; Drew *et al.*, 2000 ; reviewed by Eggers *et al.*, 2007).

Not only did storage life span decline in the 'sub-imbibed condition' relative to that of seeds stored at the shedding WCs, but fungalproliferation was exacerbated. In this respect, it is noteworthy that mild desiccation stress paradoxically stimulates germination of recalcitrant seeds before the damaging effects set in (reviewed by Eggers *et al.*, 2007).

Seeds will have been stimulated to entrain germinative metabolism sooner, when they are placed into storage after loss of a small proportion of the water originally present (reviewed by Eggers *et al.*, 2007). Hence, they become increasingly desiccation sensitive more rapidly than if not dried. This situation results in a greater water stress and thus seed debilitastion (Pammenter *et al.*, 1994), and favours more rapid fungal proliferation from seed-associated inocuulum (Calistru *et al.*, 2000; Anguelova-Merhar *et al.*, 2003; Dos Santos *et al.*, 2006).

In this regard, 'recalcitrant seeds' appear able to elaborate antifungal enzymes and other compounds (Calistru *et al.*, 2000; Anguelova-Merhar *et al.*, 2003; Dos Santos *et al.*, 2006). These defences become decreasingly effective during storage.

Fungicide treatment has been shown to be highly effective in extending storage life span of 'recalcitrant seeds' in hydrated storage (e. g. Sunilkumar and Sudharata, 1998 ; Calistru *et al.*,

2000). However, application of non-penetrative fungicides will be effective only in situations where the inoculum is primarily located on the seed surfaces !

4.4.2 Cryostorage

However effectively the storage life span of 'recalcitrant seeds' can be extended, it remains a short- to medium-term option (reviewed by Berjak and Pammenter, 2008). This situation is because of the fact that germination at the shedding WC will virtually inevitably occur!

Seedling slow growth offers an alternative to wet storage of seeds (Chin, 1996). This means of long-term conservation is less than ideal. Hence cryostorage – generally in liquid nitrogen at 196 °C or, less ideally, at below -80 ° - presently appears to offer the only option for long-term storage (reviewed by Berjak and Pammenter, 2008)!

It would be ideal if whole seeds could be cryopreserved. This event is generally not possible because 'recalcitrant seeds' of most species are large, and at high WCs when shed. As discussed above, large seeds cannot be dried rapidly, and slow dehydration to WCs commensurate with efficient cooling or freezing is lethal.

WC must be reduced to a level obviating lethal ice crystallisation during cooling for survival at ctyogenic temperatures. Successful cryopreservation of small 'non-orthodox seeds' has been achieved in caes where desiccation could be achieved rapidly (e. g. *Azadiachta indica* [Berjak and Dumet, 1996], *Warburgia salutaris* [Kioko *et al.*, 1999, 2003] and *Wasubia japonica* [Potts and Lumpkin, 2000]).

If whole seeds are optimally warmed or thawed and imbibed or rehydrated after retrieval from cryostorage, seedlings should, theoretically, be able to be generated in a greenhouse without an intervening *in vitro* stage. However, 'recalcitrant seeds' are far too large in the great majority of cases necessitating the use of the excised embryonic axes as explants for cryopresearvation.

5. Concluding remarks, future perspectives and prospects

Although a number of phenomena and mechanisms, as discussed above, have been implicated in the acquisition and maintenance of desiccation-tolerance, it seems likely that the picture is not yet complete according to Berjak and Pammenter (2008). This situation makes unequivocal identification of the differences underlying 'recalcitrant seed' behavior presently unattainable.

Similarly, it is not yet possible to present a coherent view integrating the control of the acquisition of desiccation-tolerance. In view of its complexity and our presently fragmentary understanding of the events at the control level, and also of the expression of the many phenomena characterising the acquisition and retention of desiccation-tolerance, it is perhaps not surprising that we have a long way to go before comprehending the basis of the 'recalcitrant condition' (Berjak and Pammenter, 2008).

The use of axes complicated the cryopreservation procedure. They may be injured on excission (e. g. Goveia *et al.*, 2004). Potentially injurious treatments are also required to eliminate seed–associated microorganism inoculum (Berjak *et al.*, 1999). In addition, the extent of 'flash-drying' and cooling rates must be determined (e. g. Wesley-Smith *et al.*,

2001a, b, 2004a, b). Furthermore, the desirability of using cryoprotectants needs to ascertained. Moreover, the *in vitro* technologyensuring excised axes will establish vigorous seedlings must be developed. Additionally, thawing and especially rehydration must be optimised (e. g. Berjak *et al.*, 1999; Berjak and Mycock, 2004). Further, the means for efficient dissemination of explants retrieved from cryostorage must be established (e. g. Peran *et al.*, 2006). A further aspect that could be profitably pursued is to induce a measure of axis desiccation and chilling tolerance prior to cryopreservation (e. g. Beardmore and Whittle, 2005).

6. Acknowledgements

The constructive and critical comments and contributions of Professor Patricia Berjak of the Plant Germplasm Conservation Research at the School of Biological and Conservation Sciences of the University of KwaZulu-Natal in Durban in South Africa are gratefully acknowledged.

7. References

Ajayi, S. A., Berjak, P., Kioko, J., Dulloo, M. E. and Vodouhe, R. S. (2006). Responses of fluted pumpkin (*Telfairia occidentalis* Hook.f.) seeds to desiccation, chilling and hydrated storage. *South African Journal of Botany* 72: 544-550.

Alpert, P. (2005). The limits and frontiers of desiccation-tolerant life. *Integrative and Comparative Biology* 45: 685-695.

Anguelova-Merhar, V. S., Calistru, C. and Berjak, P. (2003). A study of some biochemical and histopathological responses of wet-stored recalcitrant seeds of *Avicennia marina* infeceted with *Fusarium moniliforme*. *Annals of Botany* 92: 1-8.

Ashraf, M. (2010). Inducing drought tolerance in plants: recent advances. *Biotechnology Advances* 28: 169-183.

Ashraf, M. and Foolad, M. R. (2008). Role4s of glycine betaine and proline in improving plant abiotic stress resistance. *Environmental and Experimental Botany* 59: 206-216.

Ashraf, M., Atlar, H. R., Harris, P. J. C. and Kwon, T. R. (2007). Some perpective strategies for improving salt tolerance. *Advances in Agronomy* 97: 45-110.

Bailly, C. (2004). Active oxygen species and antioxidants in seed biology. *Seed Science Research* 4: 93-107.

Bartels, D. (2005). Desiccation tolerance studied in the resurrection plant *Craterostigma plantagineum*. *Integrative and Comparative Biology* 45: 696-701.

Beardmore, T. and Whittle, C. A. (2005). Induction of tolerance to desiccation and crypopreservation in silver maple (*Acer saccharinum*) embryonic axes. *Tree Physiology* 25: 965-972.

Berjak, P. (2005). Protector of the seeds: seminal reflections from southern Africa. *Science* 307: 47-49.

Berjak, P. (2006). Unifying perspectives of some mechanisms basic to desiccation tolerance across life forms. *Seed Science Research* 16: 1-15.

Berjak, P. And Dumet, D. (1996). Cryopreservation of seeds and isolated embryonic axers of neem (*Azidirachta indica*). *CryoLetters* 17: 99-104.

Berjak, P and Mycock, D. J. (2004). Calcium, with magnesium, is essential for normal seedling development from partially-dehydrated recalcitrant axes: a study on *trichilia dregeana* Sond. *Seed Science Research* 14: 217-231.

Berjak, P. and Pammenter, N. W. (1994). Recalcitrance is not an all-or-nothing situation. *Seed Science Research* 4: 263-264.

Berjak, P. and Pammenter, N. W. (1997a). Progress in the understanding and manipulation of desiccation-sensitive (recalcitrant) seeds. In: Ellis, R. M., Black, M., Murdoch, A. J. and Hong, T. D. (eds) *Basic and applied aspects of seed biology*. Kluwer Academic Press, Dordrecht. pp. 689-703.

Berjak, P. and Pammenter, N. W. (1997b). Important considerations pertaining to desiccation sensitivity of seeds. In: Edwards, D. G. W. and Naithani, S. C. (eds) *Seed and nursery technology of forest trees*. New Age International (P) Limited Publishers, New Dehli. pp. 43-69.

Berjak, P. and Pammenter, N. W. (2000). What ultrastructure has told us about recalcitrant seeds.*Revista Brasieleira de Fisiogia Vegetal* 12: 22-55.

Berjak, P. and Pammenter, N. W. (2001). Seed recalcitrance – current perspectives. *South African Journal of Botany* 67: 79-89.

Berjak, P. and Pammenter, N. W. (2004). Recalcitrant seeds. In: Benech-Arnold, R. L. and Sánchez, R. A. (Editors). *Handbook of seed physiology: application to agriculture*. Harworth Press, New York. pp. 305-345.

Berjak, P. and Pammenter, N. W. (2008). From *Avicennia* to *Zizania*: seed recalcitrance in perspective. *Annals of Botany* 101: 213-218.

Berjak, P., Dini, M. and Pammenter, N. W. (1984). Possible mechanisms underlying the differing dehydration responses in recalcitrant and orthodox seeds. Desiccation-associated subcellular changes in propagules of *Avicennia marina*. *Seed Science and Technology* 12: 365-384.

Berjak, P., Farrant, J. M., Mycock, D. J. and Pammenter, N. W. (1990). Recalcitrant (homoiohydrous) seeds: The enigma of their desiccation sensitivity. *Seed Science and Technology* 18: 297-310.

Berjak, P., Farrant, J. M. and Pammenter, N. W. (1989). The basis of recalcitrant seed behavior. In: Taylorson, R. B. (Editor). *Recent advances in the development of and germination of seeds*. Plenum Press, New York. pp. 89-108.

Berjak, P., Farrant, J. M. and Pammenter, N. W. (2007). Seed desiccation tolerance mechanisms. In: Jenks, M. A. and Wood, A. J. (Editors). *Plant desiccation tolerance*. Blackwell Publishing, Ames, Iowa. pp. 51-90 .

Berjak, P., Mycock, D. J., Wesley-Smith, J., Dumet, D. and Watt, M. P. (1996). Strategies for *in vitro* conservation of hydrated germplasm. In: Normah, N. M., Narimah, M. K. and Clyde, M. M. (Editors). *In vitro conservation of plant genetic resources*. Percetakan Watah Sdn. Bhd, Kuala Lumpur, Malaysia. pp. 19-52.

Berjak, P., Pammenter, N. W. and Vertucci, C. W. (1992). Homoiohydrous (recalcitrant) seeds: development status, desiccation sensitivity and the state of water in axes of *Landolphia kirkii* Dyer. *Planta* 186: 249-261.

Berjak, P., Vertucci, C. W. and Pammenter, N. W. (1993). Effects of developmental status and dehydration rate on characteristics of water and desiccation sensitivity in recalcitrant seeds of *Camellia sinensis*. *Seed Scoience Research* 3: 155-166.

Berjak, P., Walker, M., Watt, M. P. and Mycock, D. J. (1999). Experimental parameters underlying failure or success in plant germplasm conservation: a case study on zygotic axes of *Quercus robur* L. *CryoLetters* 20: 251-262.

Bewley, J. D. (1979). Physiological aspects of desiccation tolerance. *Annual Review of Plant Physiology* 30: 195-238.

Bewley, J. D. (1997). Seed germination and dormancy. *Plant Cell* 30: 1055-1066.

Bewley, J. D. and Black, M. (1994). *Seed physiology of development and germination, Second Edition*. Plenum Press, New York.

Bharuth, V, Berjak, P., Pammenter, N. W. and Naidoo, T. (2007). Responses to chilling of recalcitrant seeds of *Eklebergia capensis* from different provenances. Abstracts from the Fifth International Workshop on Desiccation Tolerance and Sensitivity of Seeds and Vegetative Plant Tissues. *South African Journal of Botany* 73: 498.

Billi, D. and Potts, M. (2002). Life and death of dried prokaryotes. *Research in Microbiology* 153, 7-12.

Black, M., Corbineau, F., Gee, H. and Cóme, D. (1999). Water content, raffinose and dehydrins in the induction of desiccation tolerance in immature wheat embryos. *Plant Physiology* 120: 463-471.

Blackman, S. A., Obendorf, R. L. and Lepold, A. C. (1995). Desiccation tolerance in developing soybean seeds: the role of stress proteins. *Physiologia Plantarum* 93: 630-638.

Boubriak, I., Dini, M. Berjak, P and Osborne, D. J. (2000). Desiccation and survival in the recalcitrant seeds of *Avicennia marina*: DNA replication, DNA repair and protein synthesis. *Seed Science Research* 10: 307-315.

Boudet, J., Buitink, J., Hoekstra, F. A., Rogniaux, Larré, C., Satour, P. and Leprince, O. (2006). Comparative analysis of the heat stable proteome of radicles of *Medicago truncatula* seeds during germination identifies late embryogenesis abundant proteins associated with desiccation tolerance. *Plant Physiology* 140: 1418-1436.

Bray, E. A. (1993). Molecular responses to water deficit. *Plant Physiology* 103: 1035-1040.

Browne, J., Turnacliffe, A. and Burnell, A. (2002). Anhydrobiosis: plant desiccation gene found in a nematode. *Nature* 416: 38.

Brunori, A. (1967). A relationship between DNA synthesis and water content during ripening *Vicia faba* seeds. *Caryologia* 20: 333-338.

Bryant, G., Koster, K. L. and Wolfe, J. (2001). Membrane behavior in seeds and other systems at low water content: the various effects of solutes. *Seeds Science Research* 11: 17-25.

Buitink, J. and Leprince, O. (2004). Glass formation in plant anhydrobiotes: survival in the dry state. *Cryobiology* 48: 215-218.

Buitink, J., Hoekstra, F. A. and Leprince, O. (2002). Biochemistry and biophysics of tolerance systems. In: Black, M. and Pritchard, H. W. (Editors). *Desiccation and survival in plants: drying without dying*. CABI Publishing, Wallingford, Oxford. pp. 293-318.

Biutink, J., Leger, J. J., Guisle, I., Vu, B. L., Wuillemse, S., Lamirault, G., Le Bars, A., Le Meur, N., Becker, A., Küster, H. and Leprince, O. (2006). Transcriptome profiling uncovers metabolic and regulatory processes occurring during the transition from desiccation-sensitive to desiccation-tolerant stages of *Medicago truncatula* seed. *Plant Journal* 47: 735-750.

Calistru, C., McLean, M., Pammenter, N. W. and Berjak, P. (2000). The effects of mycofloral infection on the viability and ultrastructure of wet-stored recalcitrant seeds of *Avicennia marina* (Forssk.) Vierh. *Seed Science Research* 10: 341-353.

Caprioli, M., Katholm, A. K., Melone, G., Ramlov, H., Ricci, C. and Santo, N. (2004). Trehalose in desiccated rotifers: a comparison between a bdelloid and a monogonont species. *Comparative Bichemistry and Physiology A – Molecular and Integrative Physiology* 139: 527-532.

Chien, C. T and Lin, P. (1997). Effects of harvest date on the storability of desiccation-sensitive seeds of *Machilus kusanoi* Hay. *Seed Science and Technology* 25: 361-371.

Chin, H. F. (1996). Strategies for conservation of recalcitrant species. In: Normah, M. N., Marimah, M. K. and Clyde, M. M. (Editors). *In vitro conservation of plant genetic resources*. Percetakan watan Sdn. Bdh, Kuala Lumpur, Malaysia. pp. 203-215.

Chin, H. F. and Roberts, E. H. (1980). *Recalcitrant crop seeds*. Tropical Press Sdn Bhd, Kaula Lumpur, Malaysia.

Clegg, J. S. (1986). The physical properties and metabolic status of *Artemia* cysts at low water contents: the 'water replacement hypothesis'. In: Leopold, A. C. (Editor). *Membranes, metabolism and dry organisms*. Comstock Publishing Associates, Ithaca, New York. pp. 167-187.

Clegg, J. S. (2005). Desiccation tolerance in encysted embryos of the animal extremophile, *Artemia*. *Integrative and Comparative Biology* 45: 715-724.

Close, T. J. (1996). Dehydrins: emergence of a biochemical role for a familyof plant dehydration proteins. *Physiologia Plantarum* 97: 795-803.

Close, T. J. (1997). Dehydrins: a commonality in the response plants to dehydration and low temperature. *Physiologia Plantarum* 100: 291-296.

Collada, C., Gomez, L., Casado, R. and Aragoncillo, C. (1997). Purification and *in vitro* chaperone activity of a class I small heat shock protein abundant in recalcitrant chestnut seeds. *Plant physiology* 115: 71-77.

Côme, D and Cornineau, F. (1996). Metabolic damage related to desiccation sensitivity. In: Ouédraogo, A. S., Poulsen, K. and Stibsgaard, F. (Editors). *Intermediate/recalcitrant tropical forest tree seeds*. International for Genetic Plant Resources Insitute, Rome and DANIDA, Humlebaek. pp. 107-120.

Connor, K. F. and Sowa, S. (2003). Effects of desiccation on the physiology and biochemistry of *Quercus alba* acorns. *Tree Physiology* 23: 1147-1152.

Corbineau, F. and Côme (1986a). Experiments of germination and storage of seeds of two dipterocarps: *Shorea roxburghii* and *Hopea odorata*. *Seed Science and Technology* 14: 585-591.

Corbineau, F. and Côme (1986b). Experiments on the storage of seeds and seedlings of *Symphonia lgobulifera* L. f. (Guttiferrae). *The Malaysian Forester* 49: 371-381.

Corbineau, F. and Côme (1988). Storage of recalcitrant seeds of four tropical species. *Seed Science and Technology* 16: 97-103.

Crane, J., Kovach, D., Gardner, C. and Walters, C. (2006). Triacylglycerol phase and 'intermediate' seed storage physiology: a study of *Cuphea carthagenensis*. *Planta* 223: 1081-1089.

Cuming, A. C. (1999). LEA proteins. In: Shewry and P. R., Casey, R. (Editors). *Seed proteins*. Kluwer Academic Publishers, Dordrecht, The Netherlands. Pp. 753-779.

Danthu, P., Gueye, A., Boye, A., Bauwens, D. and Sarr, A. (2000). Seed storage behavior of four Sahelian and Sudanian tree species (*Boscia senegalensis, Butyrospermum parkii, Cordyla pinnata* and *SABA senegalensis*). *Seed Science Research* 10: 183-187.

Daniels, M. J., Chrispeels, M. J. and Yeager, M. (1999). Projection structure of a plant vacuole membrame aquaporin by electron crystallography. *Journal of Molecular Biology* 294: 1337-1349.

Daws, M. I., Gaméné, C. S., Glidewell, S. M. and Pritchard, H. W. (2004a). Seed mass varioation masks a single critical water content in recealcitrant seeds. *Seed Science Research* 14: 185-195.

Daws, M. I., Lydall, E., Chmielarz, P., Leprince, O., Matthews, S., Thamos, C. A. and Pritchard, H. W. (2004b). Developmental heat sum influences recalcitrant seed traits in *Aesculus hippocastum* across Europe. *New Phytologist* 162: 157-166.

Daws, M. I., Cleland, H., Chmielarz, P., Gorian, F., Leprince, O., Mullins, C. E., Thanos, C. A., Vandvik, V. and Pritchard, H. W. (2006a). Variable desiccation tolerance in *Acer pseudolatanus* seeds in relation to developmental conditions: a case of phenotypic recalcitrance? *Functional Plant Biology* 33: 59-66.

Daws, M. I., Garwood, N. C. and Pritchard, H. W. (2006b). Prediction of desiccation sensitivity in seeds of woody species: a probabilistic model based on two seed traits in 104 species. *Annals of Botany* 97: 667-674.

Daws, M. I., Wood, C. B., Marks, T. and Pritchard, H. W. (2007). Desiccation sensitivity in the Arecaceae: correlated and frequency. Abstracts from the Fifth International Workshop on Desiccation Tolerance and Sensitivity of Seeds and Vegetative Plant Tissues. *South African Journal of Botany* 73: 483.

Dietz, K-J. (2003). Plant peroxiredoxins. *Annual review of Plant Biology* 54: 93-107.

Dos Santos, A. L. W., Wiethölter, N. El Gueddari, N. E. and Moerschbacher, B. M. (2006). Protein expression during seed development of *Araucaria angustifolia*: transient accumulation of class IV chitinases and arabinogalactan proteins. *Physiologia Plantarum* 127: 138-148.

Drew, P. J., Pammenter, N. W. and Berjak, P. (2000). 'Sub-imbibed' storage is not an option for extending longevity of recalcitrant seeds of the tropical species, *Trichilia dregeana*. *Seed Science Research* 10: 355-363.

Eggers, S., Erdey, D., Pammenter, N. W. and Berjak, P. (2007). Storage and germination responses of recalcitrant seeds subjected to mild dehydration. In: Adkins, S. (Editor). *Seed science research: advances and applications*. CABI Publishing, Wallingford, Oxford. pp. 85-92.

Eira, M. T. S., Walters, C., Caldas, L. S., Fazuoli, L. C., Sampaio, J. B. and Dias, M. C. L. L. (1999). Tolerasnce of *Coffea* spp. Seeds to desiccation and low temperature. *Revista Brasileira de Fisiologia* 11: 97-105.

Ellis, R. H. and Roberts, E. H. (1980). Improved equations for the prediction of seed longevity. *Annals of Botany* 45: 13-30.

Ellis, R. H., Hong, T. D. and Roberts, E. H. (1990a). An intermediate category of seed storage behavior? 1. Coffee. *Journal of Experimental Botany* 41: 1167-1174.

Ellis, R. H., Hong, T. D., Roberts, E. H. and Tao, K. L. (1990b). Low-moisture-content limits to relations between seed longevity and moisture. *Annals of Botany* 65: 493-504.

Erdey, D., Sershen, Pammenter, N. W. and Berjak, P. (2007). Drying out in Africa: physical and physiological seed characteristics of of selected indigenous plant species. Abstracts

from the Fifth International Workshop on Desiccation Tolerance and Sensitivity of Seeds and Vegetative Plant Tissues. *South African Journal of Botany* 73: .

Faria, J. M. R., Van Lamme, A. A. M. and Hilhorst, H. M. W. (2004). Desiccation sensitivity and cell cycle aspects in seeds of *Inga vera* subspp. *Affinis*. *Seed Science Research* 14: 165-178.

Faria, J. M. R., Buitink, J., Van Lamme, A. A. M. and Hilhorst, H. M. W. (2005). Changes in DNA and microtubules during loss and re-establishment of desiccation tolerance in geminating *Medicago truncutula* seeds. *Journal of Experimental Botany* 56: 2119-2130.

Farrant, J. M., Berjak, P. and Pammenter, N. W. (1993a). Studies on the development of the desiccation-sensitive (recalcitrant) seeds of *Avicennia marina* (Forsk.) Vierh.: the acquisition of germinability and response to storage and dehydration. *Annals of Botany* 71: 405-410.

Farrant, J. M., Pammenter, N. W. and Berjak, P. (1986). The increasing desiccation sensitivity of recalcitrant *Avicennia marina* seeds with storage time. *Physiologia Plantarum* 67: 291-298.

Farrant, J. M., Pammenter, N. W. and Berjak, P. (1988). Recalcitrance - a current assessment. *Seed Science and Technology* 16: 155-166.

Farrant, J. M., Pammenter, N. W. and Berjak, P. (1989). Germination-associated events and the desiccation sensitivity of recalcitrant seeds – a study of three unrelated species. *Planta* 178: 189-198.

Farrant, J. M., Pammenter, N. W. and Berjak, P. (1993b). Seed development in relation to desiccation tolerance: a comparison between desiccation-sensitive (recalcitrant) seeds of *Avicennia marina* and desiccation-tolerant types. *Seed Science Research* 3: 1-13.

Farrant, J. M., Pammenter, N. W., Berjak, P. and Walters, C. (1996). Presence of dehydrin-like proteins and levels of abscisic acid in recalcitrant (desiccation-sensitive) seeds may be related to habitat. *Seed Science Research* 6: 175-182.

Farrant, J. M., Pammenter, N. W., Berjak, P. and Walters, C. (1997). Subcellular organization and metabolic activity during the development of seeds that attain different levels of desiccation tolerance. *Seed Science Research* 7: 135-144.

Finch-Savage, W. E. and Blake, P. S. (1994). Indeterminate development in desiccation-sensitive seeds of *Quercus robur* L. *Seed Science Research* 4: 127-133.

Finch-Savage, W. E., Pramanik, S.. and Bewley, D. (1994). The expression of dehydrin proteins in desiccation-sensitive (recalcitrant) seeds of temperate trees. *Planta* 193: 478-485.

Flynn, S., Turner, R. M. and Stuppy, H. W. (2006). Seed Information DatABAse (SID) release7-0, http://www.kew.org/data/sidOctober.

Foyer,C. H. and Noctor, G. (2005). Oxidant and antioxidant signaling in plants: a re-evaluation of the concept of oxidative stress in a physiological context. *Plant, Cell and Environment* 28: 1056-1071.

Francini, A., Galleschi, L., Saviozzi, F., Pinzino, C., Izzo, R., Sgherri, C. and Navari-Izzo, F. (2006). Enzymatic and non-enzymatic protective mechanisms in recalcitrant seeds of *Araucaria bidwilli* subjected to desiccation. *Plant Physiology and Biochemistry* 44: 556-563.

Fridovich, I. (1998). Oxygen toxicity: a radical explanation. *Journal of Experimental Biology* 201: 1203-1209.

Galau, G. A. and Hughes, D. W. (1987). Coordinate accumulation of homologous transcripts of seven cotton *lea* gene families during embryogenesis and germination. *Developmental Biology* 123: 213-231.

Galau, G. A., Bijaisorodat, N. and Hughes, D. W. (1987). Accumulation kinetics of cotton late embryogenesis-abundant mRNAs: coordinate regulation during embryogenesis and the role of abscisic acid. *Developmental Biology* 123: 198-212.

Galau, G. A., Hughes, D. W. and Dure, L. S. III (1986). Abscisic acid induction of cloned cotton late embryogenesis-abundant (*lea*) mRNAs. *Plant Molecular Biology* 123: 198-212.

Gaméné, C. S., Pritchard, H. W. and Daws, M. I. (2004). Effects of desiccation and storage on *Vitellaria paradoxa* seed viability. In: Sacandé, M., Joker, D., Dolloo, M. E. and Thomsen, K. A. (Editors). *Comparative storage of biology of tropical tree seeds.* International Plant Genetic Resources Insitute, Rome. pp. 40-56.

Gee, O. H., Probert, R. J. and Coomber, S. A. (1994). 'Dehydrin-like' proteins and desiccation tolerance in seeds. *Seed Science Research* 4: 135-141.

Golovina, E. A. and Hoekstra, F. A. (2002). Membrane behavior as influenced by parttioning of amphiphiles during drying: a compararative study on anhydrobiotic plant systems. *Comparative Biochemistry and Physiology Part A* 131: 545-558.

Golovina, E. A., Hoekstra, F. A. and Hemminga, M. A. (1998). Drying increases intracellular partitioning of amphiphilic substances into the lipid phase: impact on membrane permeability and significance for desiccation tolerance. *Plant Physiology* 118: 975-986.

Goveia, M., Kioko, J. I. and Berjak, P. (2004). Developmental staus is a critical factor in the selection of excised recalcitrant axes as explants for cryopreservation. *Seed Science Research* 14: 241-248.

Guilloteau, M., Laloi, M., Blais, D., Crouzillat, D. and McCarthy, J. (2003). Oil bodies in *Theobroma cacao* seeds: cloning and characterization of cDNA encoding the 15.8 and 16.9 kDa oleosins. *Plant Science* 164: 597-606.

Gumede, Z., Merhar, V. and Berjak, P. (2003). Effect of desiccation on the microfilament component of the cytoskeleton in zygotic embryonic axes of *Trichilia dregeana*. In: *Proceedings of the Fourth International Workshop on Desiccation Tolerance and Sensitivity of Seeds and vegetative Plant Tissues.* Blouwwaterbaai, South Africa. p. 22.

Halliwell, B. (1987). Oxidative damage, lipid peroxidation and antioxidant protection in chloroplasts. *Chemistry and Physics of Lipids* 44: 327-340).

Halperin, S. J. and Koster, K. L. (2006). Sugar effects on membrane damage during desiccation of pea embryo protoplast. *Journal of Experimental Botany* 57: 2303-2311.

Harman, D. (1956). Ageing: a theory based on free radical and radiation chemistry. *Journal of Gerontology* 11: 298-300.

Hendry, G. A. F. (1993). Oxygen and free radical processes in seed longevity. *Seed Science Research* 3: 141-153.

Hong, T. D. and Ellis, R. H. (1990). A comparison of maturation drying, germination and desiccation tolerance between developing seeds of *Acer pseudoplatanus* L. and *Acer platanoides* L. *New Phytologist* 116: 589-596.

Hong, T. D. and Ellis, R. H. (1996). *A protocol to determine seed storage behavior.* In: Engels, J. M. N. and Toll, J. (Editors). International Plant Genetic Resources Institute, Rome.

Hong, T. D., Ellis, R. H., Astley, D., Pinnegar, A. E., Groot, S. P. C. and Kraak, H. L. (2005). Survival and vigour of ultra-dry seeds after ten years of hermetic storage. *Seed Science and Technology* 33: 449-460.

Horowicz, M. and Obendorf, R. L. (1994). Seed desiccation tolerance and storability: dependence on flatulence-producing oligosaccharides and cyclitols – review and survey. *Seed Science Research* 4: 385-405.

Illing, N., Denby, K. J., Collett, H., Shen, A. and Farrant, J. M. (2005). The signature of seeds in resurrection plants: a molecular and physiological comparison of desiccation tolerance in seeds and vegetative tissues. *Integrative and Comparative Biology* 45: 771-787.

International Plant Genetic Resources Insitute/DFSC. (2004). The desiccation and storage protocol. In: Secandé, M, Jøker, D., Dulloo, M. E. and Thomsen, K. A. (Editors). *Comparative storage biology of tropical tree species*. International Plant Genetic Resources Insitute, Rome. pp. 345-351.

Janská, A., Maršík, P., Zelenková, S. and Ovesná, J. (2010). Cold stress and acclimation – what is important for metabolic adjustment? *Plant Biology* 12: 395-405.

Johnson, K. D., Herman, E M and Chrispeels, M. J. (1989). An abundant, highly conserved tonoplast protein in seeds. *Plant Physiology* 91: 1006-1013.

Kaloyereas, S. A. (1958). Rancidity as a factor in loss of viability in pine and other seeds. *Journal of Oil Chemists' Society* 35: 176-179.

Kermode, A. R. (1990). Regulatory mechanismsm involved in the transition from seed development to germination. *Critical reviews in Plant Sciences* 9: 155-195.

Kermode, A. R. (1995). Regulatory mechanismsm involved in the transition from seed development to germination: interactions between the embryo and the seed environment. In: Kigel, J. and Galili, G. (Eidtors). *Seed development and germination*. Marcel Dekker, New York. pp. 273-332.

Kermode, A. R. and Finch-Savage, W. E. (2002). Desiccation sensitivity in orthodox and recalcitrant seeds in relation to development. *Desiccation and survival in plants: drying without dying.* CABI Publishing, Wallingford, Oxford. pp. 149-184.

Kioko, J. I., Berjak, P., Pritchard, H. and Daws, M. (1999). Studies on the post-shedding behavior and cryopreservation of seeds of *Warburgia salutaris*, a highly endangered medicinal plant indigenous to tropical Africa. In: Marzalina, M., Khoo, K. C., Jayanthi, N., Tsan, F. Y. and Krishanpillay, B. (Editors). *Recalcitrant seeds*. CABI Publishing, Wallingford, Oxford. pp. 365-371.

Kioko, J. I., Berjak, P., and Pammenter, N. W. (2003). Responses to dehydration and cryopreservation of seeds of *warburgia salutaris*. *South African Journal of Botany*. 69: 532-539.

Koster, K. L. (1991). Glass formation and desiccation tolerance in seeds. *Plant Physiology* 96: 302-304.

Koster, K. L. and Bryant, G. (2005). Dehydration in model membranes and protoplasts: contrasting effects at low, intermediate and high hydrations. In: Chen, T. H. H., Uemura, M. and Fujikawa, S. (Editors). *Cold hardiness in plants: molecular genetics, cell biology and physiology*. CAB International, Wallingford, Oxford. pp. 219-234.

Koster, K. L. and Leopold, A. C. (1988). Sugars and desiccation tolerance in seeds. *Plant Physiology* 88: 829-832.

Kranner, I. and Birtić, S. (2005). A modulating role for antioxidants in desiccation tolerance. *Integrative and Comparative Biology* 45: 734-740.

Kranner, I., Beckett, R. P., Wornik, S., Zorn, M., and Pfeifhofer, H. W. (2002). Revival of a resurrection plant correlates with its antioxidant status. *Plant Journal* 31: 13-24.

Kranner, I., Cram, W. J., Zorn, M., Wornik, S., Yoshimura, I., Stabenheimer, E. and Pfeifhofer, H. W. (2005). Antioxidant and photoprotection in a lichen as compared with its isolated symbiotic partners. *Proceedings of the National Academy of Sciences, USA* 102: 3141-3146.

Laloi, C., Apel, K. and Damon, A. (2004). Reactive oxygen signaling: the latest news. *Current Opinion in Plant Biology* 7: 323-328.

Leprince, O., Buitink, J. and Hoekstra, F. A. (1999). Axes and cotyledons of recalcitrant seeds of *Castanea sativa* Mill. exhibit contrasting responses of respiration to drying in relation to dersiccation sensitivity. *Journal of Experimental Botany* 50: 1515-1524.

Leprince, O., Hendry, G. A. F. and McKersie, B. D. (1993). The mechanisms of desiccation tolerance in developing seeds. *Seed Science Research* 3: 231-246.

Leprince, O., Van Aelst, A., Pritchard, H. W. and Murphy, D. J. (1998). Oleoosins prevent oil-body coalescence during seed imbibitions as suggested by alow-temperature scanning electron microscope styudy of desiccation-tolerant and sensitive seeds. *Planta* 204: 109-119.

Lapinski, J. and Turnacliffe, A. (2003). Anhydrobiosis without trehalose in bdelloid rotifers. *FEBS Letters* 533: 387-380.

Leubner-Mtzger, G. (2005). B-1,3-glucanase gene expression in low-hydrated seeds as a mechanism for dormancy release during tobacco after-ripening. *Plant Journal* 41: 133-145

Liang, Y. H. and Sun, W. (2002). Rate of dehydratyion and cumulative desiccation stress integrated to modulate desiccation tolerance of cocoa and ginkgo embryonic tissues. *Plant Physiology* 128: 1323-1331.

Lin, T.-P. and Chen, M-H. (1995). Biochemical characteristics associated with the development of the desiccation-sensitive seeds of *Machilus thunbergii* Sieb & Zucc. *Annals of Botany* 76: 381-387.

Liu, M.-S. Chang, C.-Y. and Lin, T.-P. (2006). Comparison of phospholipids and their fatty acids in recalcitrant and orthodox seeds. *Seed Science and Technology* 34: 443-452.

Mansour, M. M. F. (2000). Nitrogen containing compounds and adaptation of plants to salinity stress. *Biology of Plants* 43: 491-500.

Mariaux, J. B., Bockel, C., Salamini, F. and Bartels, D. (1998). Desiccation- and ABA – responsive genes encoding major intrinsic proteins (MIPs) from the resurrection plant *Craterostigma plantagineum*. *Plant Molecular Biology* 38: 1089-1099.

Maurel, C., Chrispeels, M., Lurin, C., tacnet, F., Geelen, D., Ripoche, P. and Guern, J. (1997). Function and regulation of seed aquaporins. *Journal of Experimental Botany* 48: 421-430.

Mohanty, A., Kathuria, H., Ferjani, A., Sakamoto, A., Mohanty P., Murata, N. and Tyagi, A. (2002). Trasnsgenics of an elite indica rice variety Pusa Bumati 1 harbouring the *coda* gene are highly tolerant to salt stress. *Theoretical and Applied Genetics* 106: 51-57.

Mycock, D. J., Berjak, P. and Finch-Savage, W. E. (2000). Effects of desiccation on subcellular matrix of the embryonic axes of *Quercus robur*. In: Black, M., Bradford, K. J. and

Vázquez-Ramos, J. (Editors). *Seed biology: advances and applications.* CABI Publishing, Wallingford, Oxford. pp. 197-183.

Neya, O., Golovina. E. A., Nijsse, J. and Hoekstra, F. A. (2004). Ageing increases the sensitivity of neem (*Azadirachta indica*) seeds to imbibitional stress. *Seed Science Research* 14: 205-217.

Nkang, A., Omokaro, D., Egbe, A. and Amanke, G. (2003). Variations in fatty acid proportions during desiccation of *Telfairia occidentalis* seeds harvested at physiological and agronomic maturity. *African Journal of Biotechnology* 2: 33-39.

Ntuli, T. M. (2011a). Some biochemical studies on respiratory metabolism. In: Ntuli T. M. *Aspects of the influence of drying rate and wet storage on the physiology and biochemistry of embryonic axes from desiccation-sensitive seeds.* VBM Verlag Publishing, Beau-Bassin, Mauritius. pp. 95-114.

Ntuli, T. M. (2011b). The role of free radical processes in seed deterioration. In: Ntuli T. M. *Aspects of the influence of drying rate and wet storage on the physiology and biochemistry of embryonic axes from desiccation-sensitive seeds.* VBM Verlag Publishing, Beau-Bassin, Mauritius. pp. 115-145.

Ntuli, T. M. (2011c). Aspects of water relations during desiccation and wet storage. In: Ntuli T. M. *Aspects of the influence of drying rate and wet storage on the physiology and biochemistry of embryonic axes from desiccation-sensitive seeds.* VBM Verlag Publishing, Beau-Bassin, Mauritius. pp. 56-73.

Ntuli, T. M. (2011d). Biochemical, biophysical and physiological assessment of seed quality. In: Ntuli T. M. *Aspects of the influence of drying rate and wet storage on the physiology and biochemistry of embryonic axes from desiccation-sensitive seeds.* VBM Verlag Publishing, Beau-Bassin, Mauritius. pp. 74-94.

Ntuli, T. M., and Pammenter, N. W. (2009). Dehydration kinetics of embryonic axes from desiccation-sensitive seeds: an assessment of descriptive models *Journal of Integrative Plant Biology* 51: 1002-1007.

Ntuli, T. M., Berjak, P., Pammenter, N. W. and Smith, M. T. (1997). Effects of temperature on the desiccation responses of seeds of *Zizania palustris. Seed Science Research* 7: 145-160.

Ntuli, T. M., Finch-Savage, W. E., Berjak, P. and Pammenter, N. W. (2011). Increased drying rate lowers the critical water content for survival in embryonic axes of English oak (*Quercus robur* L) seeds. *Journal of Integrative Plant Biology* 53: 270-280.

Obendorf, R. L. (1997). Oligosaccharides and galactosyl cyclitols in seed desiccation tolerance. *Seed Science Research* 7: 61-74.

Oliver, A. E., Leprince, O.,Wolkers, W. F., Hinchta, D. K., Heyer, A/. G. and Crowe, J. H. (2001). Non-disaccharide-based mechanisms of protection during drying. *Cryobiology* 43: 151-167.

Osborne, D. J. (1983). Biochemical control of systems operating in the early hours of germination. *Canadian Journal of Botany* 61: 3568-3577.

Owttrim, C. W. (2006). RNA helicases and abiotic stress. *Nucleic Acid Research* 34: 3220-3330.

Pammenter, N. W. and Berjak, P. (1999). A review of recalcitrant seed physiology in relation to desiccation-tolerance mechanisms. *Seed Science Research* 9: 13-37.

Pammenter, N. W., Berjak, P., Farrant, J. M., Smith, M. T. and Ross, G. (1994). Why do stored hydrated recalcitrant seeds die? *Seed Science Research* 4: 187-191.

Pammenter, N. W., Greggains, V., Kioko, J. I., Wesley-Smith, J., Berjak, P. and Finch-Savage, W. E. (1998). Effects of differential drying rates on viability retention of recalcitrant seeds of *Ekebergia capansis*. *Seed Science Research* 8: 463-471.

Pammenter, N. W., Naidoo, S. and Berjak, P. (2002a). Desiccation rate, desiccation response and damage accumulation: can desiccation sensitivity be quantified? In: Nicolás, N. Bradford, K. J., Cóme, D. and Pritchard, H. W. (Editors). *The biology of seeds – recent advances*. CABI Publishing, Wallingford, Oxford. pp. 319-325.

Pammenter, N. W., Berjak, P., Wesley-Smith, J. and Vander Willigen, C. (2002b). Experimental aspects of drying and recovery. In: Black, M. and Pritchard, H. W. (Editors). *Desiccation and survival in plants: drying without dying*. CABI Publishing, Wallingford, Oxford. pp. 93-110.

Parsegian, V. A. (2002). Protein-water interactions. *International Review of Cytology* 215: 1-31.

Perán, R., Berjak, P., Pammenter, N. W. and Kioko, J. I. (2006). Cryopreservation, encapsulation and promotion of shoot production of embnryonic axes of a recalcitrant species, *Ekebergia capansis* Sparrm. *CryoLetters* 27:1-12.

Potts, S. E. and Lumpkin, T. A. S. (2000). Cryopreservation of *Wasubia* spp. seeds. *CryoLetters* 18: 1-12.

Pritchard, H. W., Tompsett, P. B., Manger, K. and Smidt, W. J. (1995). The effect of moisture content on the low temperature responses of *Araucaria hunsteneii* seed and embryos. *Annals of Botany* 76: 79-88.

Pritchard, H. W., Wood, C. B., Hodges, S. and Vautier, H. J. (2004a). 100-seed test for desiccation tolerance and germination: a case study on 8 tropical palm species. *Seed Science and Technology* 32: 393-403.

Pritchard, H. W., Daws, M. I., Fletcher, B. J., Gaméné, Msanga, H. P. and Omondi, W. (2004b). Ecological correlates of seed desiccation tolerance in tropical African dryland trees. *American Journal of Botany* 91: 393-403.

Proctor, M. C. F. and Pence, V. C. (2002). Vegetative tissues: bryophytes, vascular resurrection plants and vegetative propagules. In: Black, M. and Pritchard, H. W. (Editors). *Desiccation and survival in plants: drying without dying*. Wallingford, CABI Publishing. pp. 207-237.

Ratajczak, E. and Pukacka, S. (2006). Changes in ascorbate-glutathione system during storage of recalcitrant seeds of *Acer saccharium* L. *Acta Societatis Botanicorum Poloniae* 75: 23-27.

Rinne, P. L H., Kaikuranta, P. L. M., Van Der Plas, L. H. W. and Van Der Schoot, C. (1999). Dehydrins in cold-acclimated apices of binch (*Betula pubescens* Ehrh.): production, localization and potential role in rescuing enzyme function during dehydration. *Planta* 209: 377-388.

Roberts, E. H. (1973). Predicting the storage life of seeds. *Seed Science and Technology* 1: 499-514.

Rogerson, N. E. and Matthews, S. (1977). Respiratory and carbohydrate changes in developing pea (*Pisum sativum*) seeds in relation to their ability to withstand desiccation. *Journal of Experimental Botany* 28: 304-313.

Sacandé, M., Jøker, D., Dulloo. M. E., Thomsen, K. A. (Editors) (2004). *Comparative storage biology of tropical tree seeds*. International Plant Genetic Resources Insitute, Rome.

Sen, S. and Osborne, D. J. (1974). Germination of rye embryos following hydrationdehydration treatments: enhancement of protein and RNA synthesis and earlier induction of DNA replication. *Journal of Experimental Botany* 25: 1010-1019.

Serraj, R. and Sinclair, T. R. (2002). Osmolyte accumulation:can it really increase crop yield under drought conditions. *Plant, Cerll and Environment* 25: 333-341.

Solomon, A., Solomn, R., Paperna, I. and Glazer, I. (2000). Desiccation stress of entomopathogenic nematodes induces the accumulation of a novel heat-stable protein. *Parasitology* 121: 409-416.

Song, S.-Q., Berjak, P. and Pammenter, N. (2004). Desiccation sensivity of *Trichilia dregeana* axes and antioxidant role of ascorbic acid. *Acta Botanica Sinica* 46: 803-810.

Stacey, R. A. P., Nordeng, T. W., Culiñáez-Maciá, F. A. and Aalen, R. B. (1999). The dormancy-related peroxiredoxins antioxidant PER1 is located to the nucleus in barley embyos and aleurone. *Plant Journal* 19: 1-8.

Sun, W. Q. (1999). Desiccation sensitivity of recalcitrant seeds and germinated orthodox seeds: can germinated orthodox seeds serve as model systems for studies on recalcitrance? In: Marzalina, M. Khoo, K. C., Jayanthi, N., Tsan, F. Y and Krishnapillay, B. (Editors). *Recalcitrant seeds*. Forest Research Insitute, Kuala Lumpur, Malaysia. pp 29-42.

Sunilkumar, K. K. and Sudhakara, K. (1998). Effects of temperature, media and fungicides on the storage behavior of *Hopea parviflora* seeds. *Seed Science and Technology* 26: 781-797.

Suzuki, N. and Mittler, R. (2006). Reactive oxygen species and temperature stresses: a delicate balance between signaling and destruction. *Physiologia Plantarum* 126: 45-51.

Timasheff, S. N. (1982). Preferential interactions in protein-water co-solvent systems. In: Franks, F. and Mathias, S. F. (Editors). *Biophysics of water*. Marcel Dekker, New York. pp. 70-72.

To, A., Valon, C., Savino, G., Jocylyn, G., Devic, M. and Parcy, F. (2006). A network of local and redundant gene regulation governs *Arabidopsis* seed maturation. *Plant Cell* 18: 1642-1651.

Tommasi, F., Paciollla, C., De Pihno, M. C. and De Gara, L. (2006). Effects of storage temperature on viability, germination and antioxidant metabolism in *Ginkgo biloba* L. seeds. *Plant Physiology and Biochemistry* 44: 359-368.

Tompsett, P. B. and Pritchard, H. W. (1993). Water status changes during development in relation to the germination and desiccation tolerance of *Aesculus hippocastum* L. seeds. *Annals of Botany* 71: 107-116.

Vashisht, A. A. and Tuteja, N. (2006). Stree response to DEAD-box helicases: a new pathway to engineer plant stress tolerance.*Journal of Photochemistry and Photobiology B Biology* 84: 150-160.

Vertucci, C. W. and Farrant, J. M. (1995). Acquisition and loss of desiccation-tolerance. In: Kigel, J. and Galili, G. (Editors). *Seed development and germination*. Marcel Dekker Inc., New York. pp. 237-271.

Vincente-Carbajosa, J. and Carbanero, P. (2005). Seed maturation: developning an intrusive phase to accomplish a quiescent state. *International Journal of Devlopment* 49: 645-651.

Walters, C. (1998). Understanding the mechanisms and kinetics of seed aging. *Seed Science Research* 8: 223-244.

Walters, C. and Engels, J. (1998). The effects of storing seeds under extremely dry conditions. *Seed Science Research* 8 (Supplement Number 1): 3-8.

Walters, C., Farrant, J. M., Pammenter, N. W. and Berjak, P. (2002). Desiccation stress and damage. In: Black, M. and Pritchard, H. (eds) *Desiccation and survival in plants: drying without dying*. CAB International, Wallingford. pp. 263-291.

Walters, C., Hill, L. M. and Wheeler, L. M. (2005a). Drying while dry: kinetics and mechanisms of deterioration in desiccated organisms. *Integrative and Comparative Biology* 45: 751-758.

Walters, C., Pammenter, N. W., Berjak, P. and Crane, J. (2001). Desiccation damage, accelerated ageing and respiration in desiccation-tolerant and sensitive seeds. *Seed Science Research* 11: 135-148.

Walters, C., Wheeler, L. and Grotenhuis, J. M. (2005b). Longevity of seeds stored in a genebank: species characteristics. *Seed Science Research* 15: 1-20.

Wesley-Smith, J., Pammenter, N. W., Berjak, P. and Walters, C. (2001a). The effects of two drying rates on the desiccation tolerance of embryonic axes of recalcitrant jackfruit (*Artocarpus heterophyllus* Lumk.) seeds. *Annals of Botany* 88: 653-664.

Wesley-Smith, J., Walters, C., Pammenter, N. W. and Berjak, P. (2001b). Interactions of water content, rapid (non-equilibrium) cooling to -196 °C and survival of embryonic axesof *Aesculus hippocastum* L. seeds. *Cryobiology* 42: 196-206.

Wesley-Smith, J., Walters, C., Pammenter, N. W. and Berjak, P. (2004a). Non-equilibrium cooling of *Poncirus triofoliata* L. embryonic axes at various water contents. *CyroLetters* 25: 121-128.

Wesley-Smith, J., Walters, C., Pammenter, N. W. and Berjak, P. (2004b). The influence of water content, cooling and warming rates upon survival of embryonic axes of *Poncirus triofoliata. CyroLetters* 25: 129-138.

Williams, R. J. and Leopold, A. C. (1989). The glassy state in corn embryos. *Plant Physiology* 89: 977-981.

Wolkers, W. F., McCready, S., Brandt, W. F., Lindsey, G. G. and Hoekstra, F. A. (2001). Isolation and characterization of a D-7 LEA protein in pollen that stabilizes glasses *in vitro. Biochimica et Biophysica* 1544: 196-206.

Woodenberg W., Erdey, D., Pammenter, N. W. and Berjak, P. (2007). Post-shedding seed behaviour of selected *Encephalartos* species. Abstracts from the Fifth International Workshop on Desiccation Tolerance and Sensitivity of Seeds and Vegetative Plant Tissues. *South African Journal of Botany* 73: 496.

Yang, W. J., Rich, P. J., Axell, J. D., Wood, K. V., Bonham, C. C., Ejeta, G., Mickelbart, M. V., and Rhodes, D. (2003). Genotypic variation for glycine betainein sorghum. *Crop Science* 43: 162-169.

2

Flooding Stress on Plants: Anatomical, Morphological and Physiological Responses

Gustavo Gabriel Striker

IFEVA-CONICET, Faculty of Agronomy, University of Buenos Aires
Argentina

1. Introduction

Flooding is a natural disturbance affecting crop and forage production worldwide due to the detrimental effects that it provokes on most terrestrial plants (Bailey-Serres & Voesenek, 2008; Colmer & Vosenek, 2009). Over the last years, the Intergovernmental Panel on Climate Change (http://www.ippc.ch) has informed that man-induced world climate change will increase the frequency of precipitations of higher magnitude as well as tropical cyclone activity. As a result, the occurrence of flooding events on flood plains (*i.e.* lowlands) and cultivated lands is expected to be higher (Arnell & Liu, 2001). On the other hand, the increasing world population, along with the intensification of agriculture have provoked a reduction in the arable land per capita, which has decreased over the last five decades from 0.32 ha to 0.21 ha, and it is expected to be further diminished up to 0.16 ha per capita by 2030 (FAO 2006 as cited in Mancuso & Shabala, 2010). As a consequence, marginal lands are being incorporated into production to cope with the rising food demand. These issues lead to the necessity to get highly productive crops in arable lands subjected to periodic events of water excess, and to introduce new (or improved) flood-tolerant forage species in flood-prone pastures (and grasslands) devoted to livestock production. So, the understanding of plant functioning under flooding conditions is crucial in order to achieve these goals.

Soil water excess determines a severe decrease in the oxygen diffusion rate into the soil because of the 10^4 lower diffusion of gases into water with respect to air (Armstrong, 1979; Ponnamperuma, 1972; 1984). Shortly after the soil is flooded, the respiration of roots and micro-organisms depletes the remnant oxygen and the environment becomes hypoxic (*i.e.* oxygen levels limit mitochondrial respiration) and later anoxic (*i.e.* respiration is completely inhibited; Blom & Voesenek, 1996; Bailey-Serres & Voesenek, 2008; Wegner, 2010). So, the first constraint for plant growth under flooding is the immediate lack of oxygen necessary to sustain aerobic respiration of submerged tissues (Armstrong, 1979; Vartapetian & Jackson, 1997; Voesenek et al., 2004). As flooding time increases, a second problem associated with water excess appears as a result of the progressive decrease in the soil reduction-oxidation potential (redox potential) (see Fig. 1; Pezeshki & DeLaune, 1998; Pezeshki, 2001). With the reduction of the soil redox potential potentially toxic compounds appear such as sulfides, soluble Fe and Mn, ethanol, lactic acid, acetaldehyde and acetic and formic acid (Kozlowski, 1997; Fiedler et al., 2007). Therefore, the lack of oxygen and later the accumulation of some potentially toxic compounds are the major constraints that plants suffer under flooding conditions.

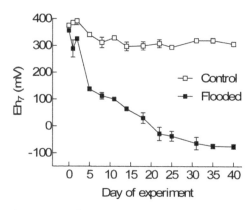

Fig. 1. Course of soil redox potential at pH 7 (EH_7) at control and flooded conditions by 40 days. [Reproduced from Striker et al. (2005) with permission from Springer].

In a broad sense, the term flooding is often used to depict different situations in which the water excess can range from water saturated soil (*i.e.* waterlogging) to deep water columns causing complete submergence of plants (Fig. 2). So, a first step is to accurately define the correct terms for each situation of water excess. Waterlogging corresponds to the full saturation of the soil pores with water, and with a very thin – or even without - a layer of water above the soil surface. Hence, under waterlogged conditions, only the root system of plant is under the anaerobic conditions imposed by the lack of oxygen, while the shoot is under atmospheric normal conditions. Flooding is the situation in which there is a water layer above the soil surface. This water layer can be shallow or deep, so that it can provoke partial or complete submergence of plants. It should be noted that, at the same water depth, the degree of plant submergence will depend on the developmental stage (*eg.* seedlings *vs.* adult plants) and plant growth habit (*eg.* creeping plant growth *vs.* erect plant growth), among other traits influencing plant height. Under partial submergence conditions, plants have a portion of their shoots underwater, besides having their roots completely immersed in water-saturated soil. Under complete submergence, plants confront the most stressful scenario because both, shoot and root plant compartments, are underwater, and in this case

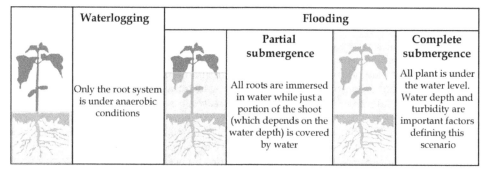

Waterlogging	Flooding		
	Partial submergence		**Complete submergence**
Only the root system is under anaerobic conditions	All roots are immersed in water while just a portion of the shoot (which depends on the water depth) is covered by water		All plant is under the water level. Water depth and turbidity are important factors defining this scenario

Fig. 2. Scheme of the different scenarios encountered by plants in front to increasing levels of water excess, ranging from waterlogging to complete submergence.

the chances to capture atmospheric oxygen and to continue with carbon fixation are restricted (but see plant strategies to deal with this stress on section 3.1). This situation is worsened in turbid water and/or with deep water columns above plants because the irradiance available to sustain underwater photosynthesis for survival is drastically reduced (Mommer et al., 2004; Colmer & Pedersen, 2008; Vashist et al., 2011).

Another crucial aspect that should be taken into account when defining 'flooding' is its duration (see Colmer & Voesenek, 2009). In this sense, flooding duration has been recognized as a major factor in determining plant survival following oxygen deprivation (Kozlowski & Pallardy, 1984; Armstrong et al., 1994; Lenssen et al., 2004). It is known that a single species of a similar age and size that is capable of surviving a short flooding period may perish if exposed to a longer one (Else et al., 1996; Crawford, 2003). In addition, a recent review of methodological aspects of flooding experiments highlighted the importance of also considering the type and age of the species tested (Striker, 2008; Fig. 3). This work showed that: (i) crop species are subjected to shorter flooding periods than non-crop species, and that (ii) seedlings of crops are exposed to even shorter periods than adult individuals; a fact that did not occur in experiments that used non-crop species (Fig. 3).

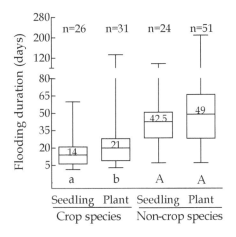

Fig. 3. Flooding duration in the experiments done on crop species and non–crop species. The end of the boxes defines the 25th and 75th percentiles, with a line at the median and error bars defining the 10th and 90th percentiles. Different letters indicate significant differences (P< 0.05) between medians based on the Mann-Whitney test. Lower case letters compare medians within the crop category while upper case letters compare medians within the non–crop category. [Adapted from Striker (2008) with permissions from Wiley-Blackwell]

2. Plant responses to partial submergence

Plants develop a suite of anatomical, morphological and physiological responses in order to deal with partial submergence imposed by flooding (Armstrong, 1979; Kozlowski & Pallardy, 1984; Vartapetian & Jackson, 1997; Striker et al., 2005; Colmer & Voesenek, 2009). The most common anatomical response is the generation of aerenchyma in tissues (Justin &

Armstrong, 1987; Seago et al., 2005), which facilitates the transport of oxygen from shoots to roots (Colmer, 2003a). At morphological level, usual responses to flooding include adventitious rooting and increases in plant height and consequently, in the proportion of biomass above water level (Naidoo & Mundree, 1993; Grimoldi et al., 1999). This also helps to facilitate the oxygenation of submerged tissues through the aerenchyma tissue (Laan et al., 1990; Colmer, 2003a). At physiological level, flooding modifies water relations and plants carbon fixation. Closing of stomata, with or without leaf dehydration, reduction of transpiration and inhibition of photosynthesis, are responses that can occur in hours or days, depending on the tolerance to flooding of each plant species (Bradford & Hsiao, 1982; Else et al., 1996; Insausti et al., 2001; Striker et al., 2005; Mollard et al., 2008; 2010). The following sections show the main plant responses at those levels associated with tolerance to flooding.

2.1 Anatomical traits of tolerance to partial submergence

Aerenchyma formation in the root cortex is the most studied plastic response to flooding (Smirnoff & Crawford, 1983; Justin & Armstrong, 1987; Colmer et al., 1998; Visser et al., 2000; McDonald et al., 2002; Evans, 2003; Grimoldi et al., 2005; Seago et al., 2005; Striker et al., 2007a). This aerenchymatic tissue provides a continuous system of interconnected aerial spaces (aerenchyma lacunae) of lower resistance for oxygen transport from aerial shoots to submerged roots, allowing root growth and soil exploration under anaerobic conditions (Armstrong, 1979; Colmer & Greenway, 2005).

The spatial arrangement of aerenchyma in the root cortex in response to flooding is variable among species (Smirnoff & Crawford, 1983; Justin & Armstrong, 1987; Visser et al., 2000; McDonald et al., 2002; Grimoldi et al., 2005; Seago et al., 2005). Different aerenchyma types arise from the combination of four general root structural types (Justin & Armstrong, 1987). Such four general root structural types - graminaceous, cyperaceous, *Apium*, and *Rumex* - have been described on the basis of the spatial arrangement of the aerenchyma tissue and the packing of the cells in the cortex (Justin & Armstrong, 1987; Seago et al., 2005). The shape of these root types resembles a bicycle wheel (graminaceous), a spider web (cyperaceous), a honeycomb (*Rumex*) and a non-organized structure with irregular aerenchyma lacunae (*Apium*) (Justin & Armstrong, 1987; Striker et al., 2007a; Fig. 4).

Three different origins of aerenchyma tissue generation have been recognized after the comprehensive review by Seago et al. (2005), namely: lysigeny, schizogeny and expansigeny. The most common is lysigeny, a process that involves the collapse and death of cells in the cortex zone, often coupled with cell separations preceding cell collapse (schizo-lysigeny). Within this aerenchyma origin, two distinct patterns leading to aerenchyma lacunae can be distinguished. The first one is called radial lysigeny, in which aerenchyma lacunae are generated by collapse of cells radially aligned in the cortex and separated by intact radial files of cells (or remnant cell walls). This type of aerenchyma is typical of many graminaceous species and resembles the shape of a bicycle wheel (Fig. 4a). The second one is termed tangential lysigeny, which implies cell separation and collapse in tangential sectors of the root cortex with intact radial files of cells, so that the resulting shape resembles a spider web. This aerenchyma type is typical of cyperaceous species (Fig. 4b). Sometimes, aerenchyma lacunae generated by cell lysis (lysineny) does not present a regular and easily identifiable spatial pattern. The last case is that of species showing an irregular non-organized location of aerenchyma lacunae in their root cortex (*Apium* root type; Fig. 4c). Schizogeny is the process of aerenchyma generation that involves the expansion of intercellular spaces into lacunae along radial sectors (*sensu* Seago et al., 2005). This origin often precedes the cell lysis (lysogeny) that increases the size of the lacunae generated and just in few cases, it appears as the only process generating aerenchyma. Expansigeny implies the generation of lacunae by cell division and cell enlargement but without cell death (lysigeny) or further separation (schizogeny). This type of aerenchyma generation is characterized by a honeycomb (hexagonal) appearance in root cross sections and it is called *Rumex* type (Fig 4d).

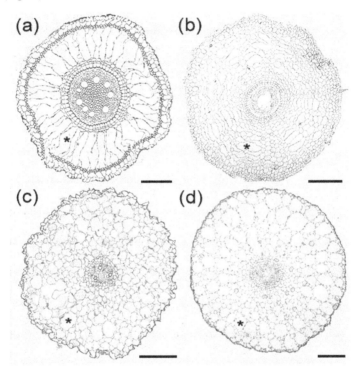

Fig. 4. Root cross sections showing the four main aerenchyma types: graminaceous in *Paspalum dilatatum* Poir. (a), cyperaceous in *Cyperus eragrostis* Lam. (b), *Apium* type in *Lotus tenuis* Waldst. & Kit. (c) and *Rumex* type in *Rumex crispus* L. (d). Asterisks indicate aerenchyma lacunae. Scale bars represent 150 μm. [Adapted from Striker et al. (2006, 2007a) with permissions from Wiley-Blackwell]

Cell death and lysis leading to aerenchyma lacunae development are attributed to low pressures of oxygen (hypoxia) and ethylene accumulation (Jackson et al., 1985; Evans, 2003; but see Visser & Bögemann, 2006). However, aerenchyma development is identical independently of being promoted by hypoxia or by ethylene (Gunawardena et al., 2001). This suggests that the cell death programme generating aerenchyma is common but it can be triggered by a variety of stimuli (Evans, 2003), also including soil mechanical impedance (Engelaar et al., 1995; Striker et al., 2006) and phosphorus deficiency (Fan et al., 2007; Postma

& Lynch, 2011). Pioneer experiments by He et al. (1994) demonstrated that the application of low oxygen or high ethylene concentrations provoked an increase of the cellulase activity in the root apex, which is likely to contribute to cell wall breaking down (final step in lysigenous aerenchyma generation) (see review by Visser & Voesenek, 2004). The events of aerenchyma formation can be described in five stages (*sensu* Evans, 2003). The first stage is the perception of hypoxia and the initiation of ethylene biosynthesis; the second stage corresponds to the perception of the signal of increasing ethylene by cells located in the root cortex (especially in the mid cortex). The third stage starts with the beginning of cell death during which ions are lost from the cell into the apoplast, the plasma membrane invagination commences and the first changes in the cell walls structure can be detected. In the fourth stage, the condensation of the chromatin to nuclear periphery is produced, cell organelles are surrounded by bilipid membranes and a marked increase in cell wall hydrolytic enzymes takes place. Finally, in the fifth stage the cell wall breakdown and cell lysis occur, and then immediately the cell content is absorbed by the surrounding cells (Schussler & Longstreth, 2000; Evans, 2003).

In herbaceous plants, oxygen transport through aerenchyma along relatively short distances (*i.e.* from shoots to roots) is mostly attributed to diffusive mechanisms. In contrast, transportation of O_2 at longer distances (metres; for instance along flooded rhizomes) is theoretically most likely to occur by convective mechanisms (see a detailed review on this topic in Colmer, 2003a; Wegner, 2010). Diffusion of oxygen under flooding conditions is established by the generation of a longitudinal gradient towards the root apex. This gradient is produced by O_2 consumption due to respiration along the root, and by the radial oxygen loss towards the rhizosphere (hereafter referred to as ROL). Both processes act as a sink of O_2 in the waterlogged soil, determining a low oxygen concentration in the root apex and consequently the generation of the mentioned gradient. It should be noticed that a higher aerenchyma generation by lysogeny (cell death in root cortex) determines a lower respiratory demand, favouring the supply of more oxygen to the root apex, at the same time that it facilitates the O_2 transport due to the lower resistance for O_2 diffusion, associated to the bigger size of the aerenchyma lacunae. In addition, the magnitude of radial oxygen loss regulates the O_2 reaching the apex, which is expected to be low if the ROL is higher (Colmer, 2003a; Colmer & Voesenek, 2009; see further on in section 2.2.3).

2.2 Morphological responses conferring plants tolerance to partial submergence

Flooding induces morphological changes in roots and shoots. In roots, the formation of adventitious roots is highlighted as a common response of flood-tolerant species. These adventitious roots, which have high porosity, help plants to continue with water and nutrient uptake under flooding conditions, replacing in some way the functions of older root system (Kozlowski & Pallardy, 1984). It is frequent that these adventitious roots are positioned near the better-aerated soil surface. Following the review by Jackson (2004), there are three mechanisms for generating these 'replacement' root systems: (i) stimulation of the outgrowth of pre-existing root primordia in the shoot base (Jackson et al., 1981), (ii) induction of a new root system that involves initiation of root primordia and their subsequent outgrowth (Jackson & Armstrong, 1999; Shimamura et al., 2007; Fig. 5a) and (iii) placing roots at the soil surface involving the re-orientation of the root extension as seen for woody species by Pereira & Kozlowski (1977) and for herbaceous species by Gibberd et al. (2001). The two first mechanisms appear to be triggered by ethylene, which is thought to increase the sensitivity of plant tissues to auxin (Bertell et al., 1990; Liu & Reid, 1992). In *Rumex palustris,* it was established that application of exogenous ethylene stimulated the production of adventitious roots without changing the root levels of indole acetic acid (IAA, an auxin). These results indicate that adventitious rooting is due to an increased sensitivity of tissues to auxin and not due to an increase in its concentration (Visser et al., 1995; 1996). Complementarily, Mc-Donald & Visser (2003) showed for *Nicotiana tabacum* (tobacco) that the application of naphthaleneacetic acid (NPA) – an auxin transport inhibitor – to wild type plants reduced the adventitious root formation to the level of ethylene-insensitive transgenic plants. These antecedents strongly demonstrate that cooperation between both hormones is important in defining adventitious rooting (see also the review by Visser & Voesenek, 2004).

It is predictable that stress from soil flooding on roots also alters shoot morphology because of the close functional interdependence between both of them. In this way, flooded plants of tolerant species are often taller than their non-flooded counterparts (Fig. 5b) as a result of increases in the insertion angles and length of their aerial organs. These responses were well characterized in the dicotyledonous *Rumex palustris* by Cox et al. (2003; 2004) and Heydarian et al. (2010) among others. The faster response is the increase in the petiole angle, called hyponastic growth, where maximum angle (70-80°, an almost vertical position) is reached just in four hours (Cox et al., 2003). Next to the change in the insertion petiole angle, an increase in petiole length follows (Cox et al., 2003, Heydarian et al., 2010) in order to maximize the leaf area above the water level (Laan et al., 1990). Such lengthening of petioles is associated with the cell wall loosening due to an increase in the expression and action of expansins (Vriezen et al., 2000). It was proved that both the increase in petiole angle and lengthening, are well mimicked by treating plants with ethylene, so that this hormone appears to be involved in regulating those responses (Vriezen et al., 2000; Heydarian et al., 2010). In graminaceous species the morphological responses are analogous to those developed by dicots. For instance, in the grass *Paspalum dilatatum* the first morphological response to flooding is the increase in the tiller insertion angle (Insausti et al., 2001) followed by the elongation of the leaf sheaths, and lastly (but not always) elongation of leaf blades (Insausti et al., 2001; Mollard et al., 2008; 2010). The higher leaf sheath length of flooded plants is the result of a higher number of longer parenchymatic cells with respect to control plants (Insausti et al., 2001).

Another specific change at shoot level implies stem hypertrophy (Fig. 5a), which is a white spongy tissue with large volumes of intercellular gas spaces (Armstrong et al., 1994). This tissue is secondary aerenchyma that forms externally from a phellogen and is homologous to cork (Shimamura et al., 2010; Teakle et al., 2011). Its role seems to be increasing air space which allows for increased movement of gases between water and plant tissues (Teakle et al., 2011). Some species with capacity to develop stem hypertrophy are *Lythrum salicaria* (Stevens et al., 1997), *Lotus uliginosus* (James & Sprent, 1999), *L. tenuis* (Striker et al., 2005; Fig. 5a), *Glycine max* (Shimamura et al., 2010) and *Melilotus siculus* (Teakle et al., 2011). In woody plants, an important morphological trait developed by tolerant species is lenticels hypertrophy at the stem base (Kozlowski, 1997). It is supposed that these special structures, functionally analogous to hypertrophied stem tissue, allow oxygen entrance into shallow roots through aerenchyma and intercellular spaces (Kozlowski, 1997; Shimamura et al.,

2010). This idea was based on studies where the blocking of lenticels of waterlogged plants with lanolin determined a marked reduction in the root aeration, so that lenticels appeared as points of air entrance to the root system (Shimamura et al., 2003). In spite of the above-discussed, there is controversy about the function of hypertrophied lenticels because, in several cases, they tend to be more developed below water (Parelle et al., 2006). Hence, this location does not support the idea of enablers of oxygen entry toward the root system. Some authors proposed that it is more likely that lenticels may help maintain plant water status during flooding, by partially supplying water for the shoots and thus replacing the less-functional roots (Pezeshki, 1996; Folzer et al., 2006). The recovery of stomatal conductance of flooded plants matching in time with the appearance of hypertrophied lenticels supports the belief that they contribute to the plant water homeostasis under flooding conditions (Groh et al., 2002; Parent et al., 2008). Finally, it was proposed that hypertrophic lenticels may also allow dissipation of metabolically generated volatile compounds like ethanol, ethylene and acetaldehyde, although the physiological significance of this fact for plant performance and survival has not been assessed to date (Jackson, 2004).

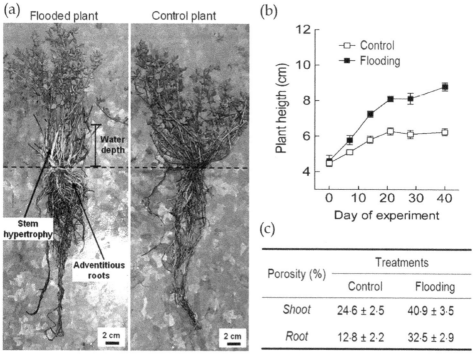

Fig. 5. Plant morphology (a) and plant height evolution (b) of *Lotus tenuis* subjected to 6 cm water depth flooding for 40 days(left photograph, note stem hypertrophy and adventitious rooting) and to control conditions (right photograph). (c) Shoot and root porosity of control and flooded plants of *L. tenuis* after 30 days of treatment. [(b) Adapted from Striker et al. (2011a) with permissions from Springer, (c) Adapted from Manzur et al. (2009) with permissions from Oxford University Press]

2.3 Physiological responses of plants to partial submergence by flooding

2.3.1 Plant water relations

In flood sensitive species like *Solanum lycopersicum, Pisum sativum, Helianthus annuus* and *Nicotiana tabacum*, a few hours after the soil becomes flooded, the water uptake by roots is reduced (Bradford & Hsiao, 1982; Jackson & Drew, 1984). Here, the reduction of water absorption under flooding is a consequence of a reduction of the root hydraulic conductivity (Else et al., 1995; Else et al., 2001; Islam & McDonald, 2004) that seems to be associated with the acidification of the cell cytoplasm and the gating of aquaporins (Tournarie-Roux et al., 2003). It appears that the excess of protons provoking such acidification, determines conformational changes of the mentioned water channels that trigger their closure (Tournarie-Roux et al., 2003; Verdoucq et al., 2008). So, the reduction of water uptake under water excess of the soil in flooding sensitive species shows the paradoxical response of wilting of leaves (Bradford & Hsiao, 1982; Else et al., 1996), as it can be seen under drought. In this type of species, unable to tolerate short-term flooding, plants die (without recovery) when the water recedes.

In flood-tolerant plants, the plant water relations during flooding can vary depending on the season of occurrence and naturally on species-specific responses (Crawford, 2003; Lenssen et al., 2004). For example, the grass *Paspalum dilatatum* and the legume *Lotus tenuis* are able to grow during periods of soil water excess in summer season (Insausti et al., 2001; Striker et al., 2006; 2007b; 2008; Mollard et al., 2008; 2010), although the impact of oxygen deprivation on the plant water status differs between species (Fig. 6). In *P. dilatatum*, flooding had no major effects on leaf water potential (Ψ_w), stomatal conductance (g_s) and transpiration rate (E) (see Figs 6a, c, e). Moreover, flooded plants registered a better plant water status than control ones on dates with higher water evaporative demand (VPD$_{air}$; Fig. 6g). By contrast, in *L. tenuis* flooding had negative effects on Ψ_w, g_s and E that increased over time and provoked 40%, 55% and 60% reductions, in relation to control plants at the end of flooding period (day 15; Figs 6b, d, f). In this sense, decreases in g_s and transpiration rate by stomatal closing in response to flooding have been proposed as a mechanism to regulate the water balance of plants and prevent leaf dehydration (Bradford & Hsiao, 1982; Ashraf, 2003; Striker et al., 2005). Here, it should be noticed that the negative effects of hypoxia in a flood tolerant species like *L. tenuis* occurred after almost a week of flooding and not in the lapse of hours, as it happened in flood-sensitive species. Importantly, when flooding was discontinued, pre-flooded plants of *L. tenuis* recovered their water status (Ψ_w) during the first five days, and showed stomatal behaviour and transpiration rates similar to control plants until the end of the recovery phase. So, plant performance during flooding alone is not conclusive for assessments of its tolerance – post-flooding recovery also needs to be appraised (Malik et al., 2002; Striker, 2008; Striker et al., 2011b). In the case of the grass *P. dilatatum*, the lack of effects of flooding on its water relations reflects its high flood-tolerance documented in several works (Loreti & Oesterheld, 1996; Insausti et al., 2001; Mollard et al., 2008; 2010; Striker et al., 2006; 2008; 2011a). The ability of this species to maintain its leaf water status similar to controls (and even better than controls, at high evaporative demand, *i.e.* high DPV$_{air}$) indicates a high capacity to continue with water uptake under flooded soil conditions. Such capacity seems to be associated with its high porous root system allowing oxygen conduction for sustaining aerobic root respiration and functionality (Insausti et al., 2001).

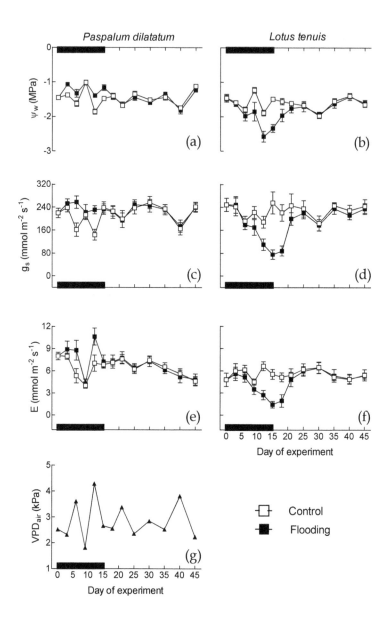

Fig. 6. Leaf water potential (Ψ_w), stomatal conductance (g_s) and transpiration rate (E) of *Paspalum dilatatum* (a, c, e) and *Lotus tenuis* (b, d, f) plants grown under different treatments: control and flooding. (g) Air vapour pressure deficit (VPD_{air}) at the moment of each measurement. Horizontal closed bars indicate the duration of the experimental flooding (15 days at 6 cm water depth). Values are means ± SE of five replicates.

2.3.2 Photosynthesis responses

A common response to flooding is the reduction of plant carbon fixation (*i.e.* rate of photosynthesis; Jackson & Drew, 1984). In the short term, photosynthesis can drop as a result of a restriction of CO_2 uptake due to stomata closing (Jackson & Hall, 1987; Huang et al., 1994; 1997; Pezeshki & DeLaune, 1998; Malik et al., 2001; Striker et al., 2005; Mollard et al., 2010). Several works have shown correlation between stomatal conductance and carbon fixation in flooded plants indicating that stomatal aperture can be a limiting factor for photosynthesis (Vu & Yelenosky, 1991; Liao & Lin, 2001; Mielke et al., 2003). Stomata closing under flooding can occur in response to leaf dehydration where the guard-cells lose their turgor (Bradford & Hsiao, 1982; Striker et al., 2007b; Fig. 6b,d), but it can also occur without noticeable changes in the leaf water potential responding to a hormonal (non-hydraulic) regulation (Jackson & Hall, 1987; Jackson, 2002; Striker et al., 2005). In the last case, the available evidence supports the idea of stomatal closure mediated by action of abscisic acid (ABA) in leaves (Else et al., 1996; Jackson et al., 2003), but not synthetized and transported from the roots, as it happens under drought stress (Davies & Zhang, 1991).

If flooding continues in time, a decrease in the photosynthetic capacity of mesophyll cells *per se* (Liao & Lin, 1994; Yordanova & Popova, 2001) leads to a further reduction of photosynthesis. Such lower photosynthetic capacity can be attributed to a (i) lower leaf chlorophyll content (Yordanova & Popova, 2001; Manzur et al., 2009; *cf.* leaf greenness of flooded *vs.* control plant in Fig. 5), (ii) a reduced activity of carboxylation enzymes, and (iii) an oxidative damage on photosystem II by reactive oxygen species (Yordanova et al., 2004). In this respect, Liao & Lin (1994) registered in *Momordica charantia* (bitter melon) a lower activation level of Rubisco (enzyme that catalyses the initial reaction during CO_2 assimilation) as flooding time increases until reaching 59% of controls values after a week of treatment. In the same experiment, these authors also registered a reduction on leaf soluble protein including Rubisco (Liao & Lin, 1994). So, both the content of Rubisco protein as well as its activation can be significantly reduced by flooding (Liao & Lin, 1994; 2001). In addition, the low photon utilization of flooded plants (Titarenko, 2000 as cited in Yordanova et al., 2004) could result in the production of reactive oxygen species (ROS) (Asada and Takahashi, 1987). The main ROS are superoxide, single oxygen, hydrogen peroxide and hydroxyl radical, which are very reactive and provoke damage to lipid membranes and proteins (see reviews by Foyer et al., 1994; Noctor & Foyer, 1998). To manage the level of ROS for protecting cells, plants have antioxidants like ascorbate, glutathione and tocopherols, and enzymes (*i.e.* peroxidases, superoxide dismutase, glutathione reductase, catalase) with ability to scavenge ROS and regenerate the antioxidants (Asada, 2006; Murchie & Niyogi, 2011). However, under flooding stress, the scavenging capacity can be over passed due to the higher production of ROS, thus generating oxidative damage on the proteins of the photosynthetic apparatus (Yordanova et al., 2004).

If it is scaled up, the negative effects of flooding on photosynthesis from the leaf level to the plant level can lead to a low growth rate in flooded plants. Such a reduction in growth, determines a low demand of triose phosphate for sucrose biosynthesis as well as a slowdown on the phloem transport of this sugar (Pezeshki, 1994; Pezeshki, 2001; Sachs & Vartapetian, 2007). Consequently, starch starts to accumulate in the chloroplasts (Wample & Davies, 1983) leading to a negative feedback on photosynthesis rate (Liao & Lin, 2001). In addition, early leaf senescence (Grassini et al., 2007) and a reduction in leaf area may also lead to a drop of carbon fixation at plant level (Striker et al., 2005). In this scenario, plants have to draw on their carbohydrate reserves in order to maintain their metabolic activity. In consequence, the level of reserve carbohydrates may be crucial in determining the tolerance to long term flooding (Schlüter & Crawford, 2001; Ram et al., 2002; Manzur et al., 2009; Striker et al., 2011a).

2.3.3 Radial oxygen loss (ROL), root apex oxygenation and root elongation

Under partial submergence (or under waterlogging), at least part of the shoots are above water and the capture of atmospheric oxygen by leaves is possible. This oxygen needs to be transported to the roots in order to avoid root anoxia. Root apex oxygenation is crucial for continuing with root elongation and soil exploration under flooding conditions (Armstrong, 1979). In plant species having (or developing) aerenchyma as a prerequisite for facilitating oxygen movement along roots, the magnitude of oxygen reaching the apex depends on the effectiveness of its longitudinal transport. When tissue respiratory demands are satisfied along the root, such effectiveness is mostly dependent on the loss of oxygen towards the rhizosphere (ROL; see review by Colmer, 2003a). The loss of oxygen from the root depends on the presence of barriers impeding its leakage towards the soil (Fig. 7). There are species possessing high aerenchyma proportion but not a barrier against ROL (or they have a slight barrier), so a considerable amount of oxygen is lost along the root, limiting the oxygen diffusion to the apex. Hence, in these species, the apex is poorly oxygenated and the root elongation is constrained (Fig. 7a). Other species, specially those inhabiting wetland sites, have barriers in the layers of the outer root cortex which prevent the loss of oxygen from the roots. In these cases, the longitudinal oxygen diffusion is enhanced, which increases the aeration of the root apex and allows root elongation and a flooded soil deeper exploration (Fig. 7b). The 'ROL barrier' can be constitutive like in *Juncus effussus* (Visser et al., 2000) or it can be induced by stagnant conditions like in *Caltha palustris* (Visser et al., 2000), *Lolium multiflorum* (McDonald et al., 2002) and some rice cultivars (Colmer et al., 1998; Colmer, 2003b). Studies on the spatial pattern of ROL along roots revealed that the barrier is preferentially located in the basal regions of the root while there is no barrier at the apex (Colmer et al., 1998; Visser et al., 2000; McDonald et al., 2002; Colmer, 2003b). So, even in species having a strong barrier against ROL, some oxygen is released through the tip zone, which generates an aerobic zone around it (Fig. 7b). It is supposed that such aerobic zone prevents the accumulation of potentially toxic compounds, like the reduced forms of iron (Fe^{2+}), manganese (Mn^{2+}) and sulfides, in the region of the sensitive root apex (Armstrong, 1979; Soukup et al., 2002; Pedersen et al., 2004; Armstrong & Armstrong, 2005).

Suberin is the most likely candidate to function as barrier to oxygen leakage (Colmer, 2003a,b; Colmer et al., 2006; Kotula et al., 2009; Shiono et al., 2011; Ranathunge et al., 2011). It should be mentioned that some authors consider that lignin cannot be discarded as part of the barrier to ROL, although evidence from De Simone et al. (2003) suggests that lignin does not accomplish such a function. Indeed, Shiono et al. (2011) indicate that suberin deposits may also be of more importance than lignin, as suberin increased prior to changes in lignin based on a detailed histochemical work. Besides, in species where the barrier against ROL is induced by stagnant conditions, there is an extra suberin deposition in the cell walls of the outer cortex, which denotes its importance in reducing ROL (Colmer, 2003b). On the other hand, it is interesting that in rice, the most studied species regarding this topic, the barrier to

ROL is triggered in a matter of hours and that the timing of generation also depends on the root length at the time of flood occurrence. Recently, Shiono et al. (2011) demonstrated that formation of the barrier to ROL commenced quickly in long adventitious roots (105 to 130 mm length) and that the barrier was completed within 24 hours. By contrast, barrier formation in short roots (65 to 90 mm length) took more than 2 days. These authors also showed that the timing of aerenchyma formation (see section 2.1) was similar between short and long roots. So, these root acclimations to deal with flooding, aerenchyma formation and the barrier to ROL, appear to be differentially regulated (see also Colmer et al., 2006).

Finally, the knowledge accomplished on the characteristics and functioning of the physical barrier against ROL has practical importance for improving the flood-tolerance of crops like wheat. This fact was recently demonstrated by Malik et al. (2011), who achieved a successful transfer of the barrier to ROL (and higher root porosity) from *Hordeum marinum* to wheat through wide hybridization and the production of *H. marinum*-wheat amphiploids.

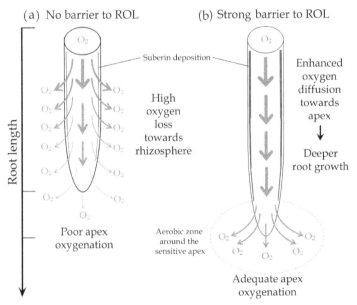

Fig. 7. Scheme showing two different patterns of radial oxygen loss (ROL) from roots. In these hypothetic examples, the root aerenchyma is considered a non-limiting factor for oxygen transport. (a) Root without barrier to ROL in the outer cortex, which loses oxygen along all positions resulting in a deficient apex oxygenation, and short roots in anoxic soils. (b) Root having a strong barrier to ROL, so that oxygen is transported efficiently to the apex allowing deeper root growth in flooded soils. The loss of oxygen is circumscribed to the apex, which generates an aerobic zone that diminishes entry of potentially toxic compounds (Fe^{2+}, Mn^{2+}, sulfides) in highly-reduced soils. The physical barrier to ROL is due to suberin deposition in the cell walls of the outer root cortex and/or the exodermis as it is indicated by red lines of different thickness between root types in (a) and (b). The thickness of the grey colour arrows indicates the amount of oxygen available. Figure re-drawn on the basis of Colmer & Voesenek (2009).

3. Plant responses to complete submergence

Complete submergence is one of the most stressful scenarios that plants can confront in environments prone to soil flooding (Blom, 1999; Mommer & Visser, 2005; Colmer & Voesenek, 2009). In addition to oxygen deficiency for roots occurring during water excess in soil, plants subjected to complete submergence are restricted from obtaining enough oxygen for sustaining tissue aeration, even though in some species, oxygen can partially be supplied by underwater photosynthesis (Mommer et al., 2004; Colmer & Pedersen, 2008; Vashist et al., 2011). As a result, aerobic metabolism for energy production shifts to the much less efficient anaerobic/fermentative pathways (Gibbs & Greenway, 2003; Voesenek et al., 2006; Kulichikhin et al., 2009). Besides, depending on the turbidity of the water, light reduction can constrain carbon gain by photosynthesis (Sand-Jensen, 1989; Colmer & Pedersen, 2008). Therefore, complete submergence can cause a drastic energy and carbohydrate crisis that can threaten plant survival (Voesenek et al., 2006; Bailey-Serres & Voesenek 2008; 2010).

According to Colmer & Voesenek (2009), this stress can be classified depending on water depth and duration of the submergence. With respect to water, shallow floods are those of less than 0.5-1 meter of water column, in which submerged plants have chances to surpass the water level if they respond elongating their shoots (Setter & Laureles, 1996; Lynn & Waldren, 2003; Hattori et al., 2007). Shallow submergence can be found in lowland flat areas of the world, as in the Flooding Pampa grasslands (Soriano, 1991), as well as in lowland rice areas. On the other hand, deep floods are those of more than 1 m of water column, in which the effort of trying to de-submerge the plant shoots is useless, because the chances to surpass the water are non-existent. In these cases, the pursued benefit of developing a shoot elongation response is not outweighed by the incurred cost, because the plant exhausts its carbohydrates reserves, dying before reaching the water surface. In contrast, plants that remain quiescent are able to succeed in front to deep submergence, surviving by using carbohydrates reserves to maintain a basal metabolism until water subsides (Schlüter & Crawford, 2001; Ram et al., 2002; Manzur et al., 2009; Striker et al., 2011b). Deep submergence can be found in areas of Asia devoted to deepwater rice cultivation, river forelands of Europe (Bloom et al., 1994), and the Amazonia of South America (Parolin, 2009). Submergence can be considered of short duration generally when it is no longer than two weeks and it occurs during flash-flooding events. If submergence period is longer than two weeks (often of a month or more), it can be regarded as of long duration (see Colmer & Voesenek, 2009 and Fig. 3 of this chapter). Although this classification can appear as arbitrary, it is useful in order to understand the strategies used by plants to deal with each combination of water depth and duration of the submergence.

3.1 Plants facing submergence. What to do, escape from water or stay quiescent?

Plants cope with complete submergence by means of one of the two major strategies reported in plant submergence responses (*sensu* Bailey-Serres & Voesenek, 2008; 2010; Fig. 8). The first is an escape strategy – called LOES: low oxygen escape syndrome – and the second is a sit-and-wait strategy – called LOQS: low oxygen quiescence syndrome (Bailey-Serres & Voesenek, 2008; 2010; Hattori et al., 2010). The LOES implies shoot elongation in order to restore leaf contact with the atmosphere, while the LOQS is based on maintaining steady energy conservation without shoot elongation (Bailey-Serres & Voesenek, 2008). It has been postulated that LOES offers plants better chances to survive under shallow long-term flooding, where shoot de-submergence is easily plausible (Fig. 8a). In contrast, LOQS is more likely to be adopted by species coping with deep short-term flooding (< 2 weeks) where shoot emergence seems to represent a higher cost of energy and might compromise eventual recovery when the water recedes (Colmer & Voesenek, 2009; Fig. 8b).

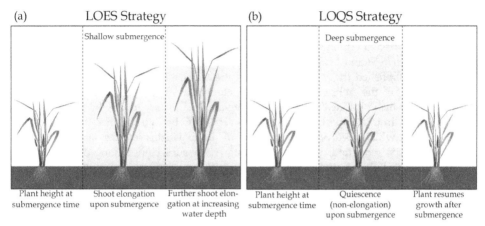

Fig. 8. Main strategies used by plants to deal with complete submergence. (a) Plant escaping from water by means of upward shoot elongation is the typical response of the 'low oxygen escape syndrome' (LOES). Plants using this strategy are able to continue elongating their shoots if the water depth increases (b) Quiescent plant under deep submergence surviving at expenses of its carbohydrates reserves depicts the 'low oxygen quiescence syndrome' (LOQS). It should be mentioned that plants escaping from water change their morphology for being taller, but their growth (in terms of biomass accumulation) is lower than that of the non-submerged plants.

Fast shoot elongation (LOES) allows plants to restore the contact of their leaves with the atmosphere under shallow submergence (Grimoldi et al., 1999; Voesenek et al., 2006; Striker et al., 2011b). So, emerged leaves can function as 'snorkels' facilitating the entrance of oxygen and ventilation of gases (ethylene, methane, CO_2) accumulated in the submerged tissues (Colmer, 2003a; Colmer & Voesenek, 2009). The shoot elongation can happen in petioles (as in *Rumex* species, see Voesenek et al., 1990; Chen et al., 2009; 2011), or internodes (as in rice, see Hattori et al., 2009; 2010) depending on the growth form of the plant. It has been established, using rice and *Rumex palustris* as model species, that one of the first signals triggering the shoot elongation is the ethylene accumulation in submerged tissues (Voesenek et al., 1993; Jackson, 2008). Ethylene accumulates in submerged tissues due to a highly restricted outward diffusion in water (Armstrong, 1979), and the upregulation of enzymes associated with its biosynthesis (*eg.* ACC synthase and ACC oxidase; Vriezen et al., 2000). Under complete submergence, concentrations of ethylene increase, which downregulates the abscisic acid levels, and upregulates those of gibberellins (Jackson, 2008). Increased gibberellins level promotes the expression of genes encoding cyclins and expansins, associated with cell division and cell extension (respectively), which lead to a fast shoot elongation underwater (Jackson, 2008). Besides, the increase in endogenous ethylene produces a lower pH in the apoplast that favours the action of expansins provoking the cell wall loosening as a necessary step that precedes cell extension (Jackson, 2008).

Quiescence syndrome of plants (LOQS) under complete submergence – of short duration but deep water column – were reported in some rice varieties (Setter & Laureles, 1996), ecotypes of *Ranunculus repens* (Lynn & Waldren, 2003), *Rumex crispus* (Voesenek et al., 1990), *Rumex acetosa* (Pierik et al., 2009), and *Lotus tenuis* (Manzur et al., 2009). In rice, pioneer studies have advanced in the understanding of this behaviour by comparing traits between tolerant accessions (*eg.* FR13A) and non-tolerant ones (*eg.* Liaogeng, IR42 and M202). Later, genetic studies have provided evidence that the shoot extension is controlled by a polygenic locus (*SUB1*) located at chromosome 9 (Xu et al., 2009). In this locus, rice has two or three genes –*SUB1A*, *SUB1B* and *SUB1C* –depending on the cultivar (or accession). Among these genes, it was established that the expression of the *SUB1A-1* is what confers submergence tolerance through the repression of shoot elongation and conservation of reserve carbohydrates, which prolongs underwater survival of plants (see review by Bailey-Serres & Voesenek, 2008). In this respect, *SUB1A-1* acts specifically by limiting the ethylene-induced shoot extension, which involves the reduction of the expansins levels related to cell wall loosening, and the reduction of sucrose and starch consumption, among other responses (Fukao et al., 2006; Fukao & Bailey-Serres, 2008; Fig. 9).

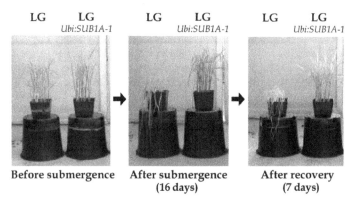

Before submergence   After submergence   After recovery
                          (16 days)   (7 days)

Fig. 9. Aspect of *Oryza sativa* plants of Liaogeng cultivar (LG, submergence intolerant) and of a transgenic line with constitutively expressed *Sub1A-1* (ubi:*SUB1A-1* with background genotype LG). Plants before submergence (left photographs), plants after 16 days of submergence (center photographs), and plants after 7 days of recovery (right photographs). It should be noticed that plants with similar genetic background (LG), but expressing Sub1A-1 are tolerant to submergence. Reproduced with modifications from Fukao & Bailey-Serres (2008) with permission granted by The National Academy of Sciences, U.S.A.

In addition to the above mentioned traits associated to LOQS, there are other ones enabling plant survival when submergence time further extends. These traits are mainly associated with the improvement of photosynthesis underwater (Mommer et al., 2004; Mommer & Visser, 2005; Colmer & Pedersen, 2008). Underwater photosynthesis is enhanced in some species due to the presence of a thin gas layer (called 'plant plastron') retained at the surface of submerged leaves, which increases the thickness of the gas-liquid interface between leaves and the surrounding water. This allows stomata to remain open when submerged, which facilitates that carbon dioxide and oxygen bypass the cuticle resistance. As a result, the levels of available CO_2 and O_2 are higher, which improve photosynthesis and oxygenation of shoots underwater (Colmer & Pedersen, 2008). Besides, some species are able to develop aquatic leaves with higher potential for carbon gain underwater. These aquatic leaves present characteristics that favour light interception underwater for improving photosynthesis, as higher specific leaf area (thin leaves), thin cuticles, thin cell walls and location of chloroplasts toward the epidermis (Mommer et al., 2004; Mommer & Visser, 2005; Bailey-Serres & Voesenek, 2010).

4. Conclusions

Floods entail different stressful conditions for plants, which mainly depend on water depth and its duration. Adaptive traits of plants enabling survival under soil waterlogging and partial submergence, are those directed to oxygenation of submerged tissues (*i.e.* parts of shoots and entire root system), and the location of leaves above water to continue with carbon fixation. Aerenchyma formation and development of adventitious roots, with barriers to radial oxygen loss, appear as the most important features facilitating longitudinal oxygen transport to sustain root aeration, and thus continue with water absorption in anaerobic soils. Both the reorientation and lengthening of shoots towards a vertical position, determine a higher proportion of leaves surpassing the water level in order to capture oxygen and continue photosynthesizing. Maintenance of stomatal conductance on mild days guarantees the uptake of CO_2 for carbon fixation, although on days of high atmospheric evaporative demand, the stomatal closure can be useful to regulate plant water homeostasis, which depends on the balance between water losses by transpiration and water uptake by roots. When water depth increases and plants are completely submerged, they can adopt two main strategies, namely LOES (low oxygen escape syndrome) and LOQS (low oxygen quiescence syndrome). The first involves the upward shoots elongation, which facilitates restoration of leaf contact with the atmosphere, and it is relevant for plants species (or ecotypes) selected in environments with shallow, prolonged floods. The second is a sit and wait strategy, where the plant remains quiescent during the submergence period by using its reserve carbohydrates conservatively for plant survival. When water subsides, plants showing LOQS resume their growth. The selection of this strategy is favoured in environments prone to deep, short floods. Future experiments assessing waterlogging and submergence responses of plants should include the combination of different flooding regimes. This would contribute to a better understanding of the costs and benefits related to particular combinations of traits conferring tolerance in variable flooding scenarios. Thus, a better comprehension of plant functioning under water excess, in a context that indicates a higher flooding occurrence during the years to come, would help to assist breeding programs as well as to define better management decisions for cultivation of crops and forage species in lands prone to flooding.

5. Acknowledgments

I thank Agustín A Grimoldi, Federico PO Mollard, and Milena E Manzur for their helpful comments and encouragement on early phases of this text preparation. I thank Beatriz Santos for her helpful revision of the text English style. This revision was funded by grants from the University of Buenos Aires (UBACyT 20020090300024) and "Agencia Nacional de Promoción Científica y Tecnológica" of Argentina (ANPCyT–PICT-2010-0205). The author is Professor of Plant Physiology at the Faculty of Agronomy of the University of Buenos Aires (Argentina) and Researcher of the Argentine National Research Council (CONICET).

6. References

Armstrong, W. (1979). Aeration in higher plants. *Advances in Botanical Research* 7, 225-332.

Armstrong, J. & Armstrong, W. (2005). Rice: sulphide-induced barriers to root radial oxygen loss, Fe^{2+} and water uptake, and lateral root emergence. *Annals of Botany* 96, 625-638.

Armstrong, W., Strange, M.E., Cringle, S. & Beckett, P.M. (1994). Microelectrode and modelling study of oxygen distribution in roots. *Annals of Botany* 74, 287-299.

Arnell, N. & Liu, C. (2001). Climatic Change 2001: hydrology and water resources. Report from the Intergovernmental Panel on Climate Change. http://www.ipcc.ch/ [Verified 28 July 2011].

Asada, K. (2006). Production and scavenging of reactive oxygen species in chloroplasts and their functions. *Plant Physiology* 141, 391-396.

Asada, K. & Takahashi, M. (1987). Production and scavenging of active oxygen in chloroplasts. *In* Photoinhibition, DJ Kyle, CB Osmond, CJ Arntzen (eds). pp 227-287. Elsevier, Amsterdam.

Ashraf, M. (2003). Relationships between leaf gas exchange characteristics and growth of differently adapted populations of Blue panicgrass (*Panicum antidotale* Retz.) under salinity or waterlogging. *Plant Science* 165, 69-75.

Bailey-Serres, J. & Voesenek L.A.C.J. (2008) Flooding stress: acclimations and genetic diversity. *Annual Review of Plant Biology* 59, 313-339.

Bailey-Serres, J. & Voesenek L.A.C.J. (2010). Life in the balance: a signaling network controlling survival of flooding. *Current Opinion in Plant Biology* 13, 489-494.

Bertell, G., Bolander, E. & Eliasson, L. (1990). Factors increasing ethylene production enhance the sensitivity of root growth to auxins. *Physiologia Plantarum* 79, 255-258.

Blom, C. W.P.M., Voesenek, L. A.C.J., Banga, M., Engelaar, W.M.H.G., Rijnders, J. H.G.M., Van De Steeg, H. M. & Visser, E. J.W. (1994). Physiological ecology of riverside species: adaptive responses of plants to submergence. *Annals of Botany* 74, 253-263.

Blom, C.W.P.M. & Voesenek, L.A.C.J. (1996). Flooding: the survival strategies of plants. *Trends in Ecology & Evolution* 11, 290-295.

Bradford, K.J. & Hsiao, T.C. (1982). Stomatal behavior and water relations of waterlogged tomato plants. *Plant Physiology* 70, 1508-1513.

Chen, X., Huber, H., de Kroon, H., Peeters, A.J.M., Poorter, H., Voesenek, L.A.C.J. & Visser E.J.W. (2009) Intraspecific variation in the magnitude and pattern of flooding-induced shoot elongation in *Rumex palustris*. *Annals of Botany* 104, 1057-1067.

Chen, X., Visser, E.J.W., de Kroon, H., Pierik, R., Voesenek, L.A.C.J., & Huber H. (2011). Fitness consequences of natural variation in flooding-induced shoot elongation in *Rumex palustris*. *New Phytologist* 190, 409-420.

Colmer, T.D. (2003a). Long-distance transport of gases in plants: a perspective on internal aeration and radial oxygen loss from roots. *Plant, Cell & Environment* 26, 17-36.

Colmer, T.D. (2003b). Aerenchyma and an inducible barrier to radial oxygen loss facilitate root aeration in upland, paddy and deepwater rice (*Oryza sativa* L.). *Annals of Botany* 91, 301-309.

Colmer, T.D., & Greenway, H. (2005). Oxygen transport, respiration, and anaerobic carbohydrate catabolism in roots in flooded soils. *In* Plant respiration: from cell to ecosystem. H Lambers, M Rivas-Carbo (eds). Pp. 137-158. Springer, The Netherlands.

Colmer, T.D. & Pedersen, O. (2008). Underwater photosynthesis and respiration in leaves of submerged wetland plants: gas films improve CO_2 and O_2 exchange. *New Phytologist* 177, 918–926.

Colmer, T.D. & Voesenek L.A.C.J. (2009). Flooding tolerance: suites of plant traits in variable environments. *Functional Plant Biology* 36, 665–681.

Colmer, T.D., Cox, M.C.H. & Voesenek, L.A.C.J. (2006). Root aeration in rice (*Oryza sativa* L.): evaluation of oxygen, carbon dioxide, and ethylene as possible regulators of root acclimatizations. *New Phytologist* 170, 767–778.

Colmer, T.D., Gibberd, M.R., Wiengweera, A. & Tinh, T.K. (1998). The barrier to radial oxygen loss from roots of rice (*Oryza sativa* L.) is induced by growth in stagnant solution. *Journal of Experimental Botany* 49, 1431–1436.

Cox, M.C.H., Benschop, J.J., Vreeburg, R.A.M., Wagemaker, C.A.M., Moritz, T., Peeters, A.J.M. & Voesenek, L.A.C.J. (2004). The roles of ethylene, auxin, abscisic acid, and gibberellin in the hyponastic growth of submerged *Rumex palustris* petioles. *Plant Physiology* 136, 2948–2960.

Cox, M.C.H., Millenaar, F.F., van Berkel, Y.E.M., Peeters, A.J.M. & Voesenek, L.A.C.J. (2003). Plant Movement. Submergence–induced petiole elongation in *Rumex palustris* depends on hyponastic growth. *Plant Physiology* 132: 282–291.

Crawford, R.M.M. (2003). Seasonal differences in plant responses to flooding and anoxia. *Canadian Journal of Botany* 81, 1224–1246.

Davies, W.J. & Zhang, J. (1991). Root signals and the regulation of growth and development of plants in drying soil. *Annual Review of Plant Physiology and Plant Molecular Biology* 42, 55–76.

De Simone, O., Haase, K., Müller, E., Junk, W., Hartmann, K., Schreiber, L. & Schmidt W. (2003). Apoplasmic barriers and oxygen transport properties of hypodermal cell walls in roots from four Amazonian tree species. *Plant Physiology* 132, 206–217

Else, M.A., Coupland, D., Dutton, L. & Jackson M.B. (2001). Decreased root hydraulic conductivity reduces leaf water potential, initiates stomatal closure and slows leaf expansion in flooded plants of castor oil (*Ricinus communis*) despite diminished delivery of ABA from the roots to the shoots in xylem sap. *Physiologia Plantarum* 111, 46–54.

Else, M.A., Davies W.J., Malone, M. & Jackson M.B. (1995). A negative hydraulic message from oxygen-deficient roots of tomato plants? Influence of soil flooding on leaf water potential, leaf expansion, and synchrony between stomatal conductance and root hydraulic conductivity. *Plant Physiology* 109, 1017–1024.

Else, M.A., Tiekstra, A.E., Croker, S.J., Davies, W.J. & Jackson, M.B. (1996). Stomatal closure in flooded tomato plants involves abscisic acid and a chemically unidentified anti–transpirant in xylem sap. *Plant Physiology* 112, 239–247.

Engelaar, W.M.H.G., Jacobs, M.H.H.E. & Blom, C.W.P.M. (1993). Root growth of *Rumex* and *Plantago* species in compacted and waterlogged soils. *Acta Botanica Neerlandica* 42, 23–35.

Evans, D. E. (2003). Aerenchyma formation. *New Phytologist* 161, 35–49.

Fan, M., Bai, R., Zhao, X. & Zhang, J. (2007). Aerenchyma formed under phosphorus deficiency contributes to the reduced root hydraulic conductivity in maize roots. *Journal of Integrative Plant Biology* 49, 598–604.

Fiedler, S., Vepraskas, M. J. & Richardson J.L. (2007). Soil redox potential: importance, field measurements, and observations. *Advances in Agronomy* 94, 2–56.

Folzer, H., Dat, J.F., Capelli, N., Rieffel, D. & Badot, P.M. (2006). Response of sessile oak seedlings to flooding: an integrated study. *Tree Physiology* 26, 759–766.

Foyer, C.H., Descourvieres, P. & Kunert, K.J. (1994). Protection against oxygen radicals: an important defense mechanism studied in transgenic plant. *Plant, Cell & Environment* 17, 507–523.

Fukao, T., Xu, K., Ronald, P.C. & Bailey-Serres J.A. (2006). Variable cluster of ethylene response factor–like genes regulates metabolic and developmental acclimation responses to submergence in rice. *Plant Cell* 18, 2021–2034.

Fukao, T. & Bailey-Serres J.A. (2008). Submergence tolerance conferred by *Sub1A* is mediated by SLR1 and SLRL1 restriction of gibberellin responses in rice. *Proceedings of the National Academy of Sciences, USA* 105, 16814–16819.

Gibberd, M.R., Gray, J.D., Cocks, P.S. & Colmer, T.D. (2001). Waterlogging tolerance among a diverse range of *Trifolium* accessions is related to root porosity, lateral root formation and aerotropic rooting. *Annals of Botany* 88, 579–589.

Gibbs, J. & Greenway, H. (2003). Mechanisms of anoxia tolerance in plants. I. Growth, survival and anaerobic catabolism. *Functional Plant Biology* 30, 1–47.

Grassini, P., Indaco, G.V., López Pereira, M., Hall, A.J. & Trápani, N. (2007). Responses to short-term waterlogging during grain filling in sunflower. *Field Crops Research* 101, 352–363.

Grimoldi, A.A., Insausti, P., Roitman, G.G. & Soriano, A. (1999). Responses to flooding intensity in *Leontodon taraxacoides*. *New Phytologist* 141, 119–128.

Grimoldi, A.A., Insausti, P., Vasellati, V. & Striker, G.G. (2005) Constitutive and plastic root traits and their role in differential tolerance to soil flooding among coexisting species of a lowland grassland. *International Journal of Plant Sciences* 166, 805–813.

Groh, B., Hübner, C. & Lendzian, K.J. (2002). Water and oxygen permeance of phellems isolated from trees: the role of waxes and lenticels. *Planta* 215, 794–801.

Gunawardena, A.H.L.A., Pearce, D.M., Jackson, M.B., Hawes, C.R. & Evans, D.E. (2001). Characterisation of programmed cell death during aerenchyma formation induced by ethylene or hypoxia in roots of maize (*Zea mays* L.). *Planta* 212, 205–214.

Hattori, Y., Nagai, K., Furukawa, S., Song, X.J., Kawano, R., Sakakibara, H., Wu, J., Matsumoto, T., Yoshimura, A., Kitano, H., Matsuoka, M., Mori, H. & Ashikari, M. (2009). The ethylene response factors *SNORKEL1* and *SNORKEL2* allow rice to adapt to deep water. *Nature* 460, 1026–1030.

Hattori, Y., Nagai, K. & Ashikari, M. (2010). Rice growth adapting to deep water. *Current Opinion in Plant Biology* 14, 1–6.

He, C.J., Drew, M.C. & Morgan, P.W. (1994). Induction of enzymes associated with lysigenous earenchyma formation in roots of *Zea mays* during hypoxia or nitrogen starvation. *Plant Physiology* 105, 861–865.

Heydarian, Z., Sasidharan, R., Cox, M.C.H., Pierik, R., Voesenek, L.A.C.J. & Peeters, A.J.M. (2010). A kinetic analysis of hyponastic growth and petiole elongation upon ethylene exposure in *Rumex palustris*. *Annals of Botany* 106, 429–435.

Huang, B., Johnson, J.W., Nesmith, D.S. & Bridges, D.C. (1994). Growth, physiological and anatomical responses of two wheat genotypes to waterlogging and nutrient supply. *Journal of Experimental Botany* 45, 193–202.

Huang, B., Johnson, J.W., Box, J.E. & NeSmith, D.S. (1997). Root characteristics and hormone activity of wheat in response to hypoxia and ethylene. *Crop Science* 37, 812–818.

Insausti, P., Grimoldi, A.A., Chaneton, E.J. & Vasellati, V. (2001). Flooding induces a suite of adaptive plastic responses in the grass *Paspalum dilatatum*. *New Phytologist* 152, 291–299.

Islam, M.A. & Macdonald, S.E. (2004). Ecophysiological adaptations of black spruce (*Picea mariana*) and tamarack (*Larix laricina*) seedlings to flooding. *Trees* 18, 35–42.

Jackson, M.B. (2002). Long-distance signalling from roots to shoots assessed: the flooding story. *Journal of Experimental Botany* 53, 175–181.

Jackson, M.B. (2004). The impact of flooding stress on plants and crops. http://www.plantstress.com/Articles/waterlogging_i/waterlog_i.htm [Verified 28 July 2011].

Jackson, M.B. (2008). Ethylene-promoted elongation: an adaptation to submergence stress. *Annals of Botany* 101, 229–248.

Jackson, M.B. & Armstrong, W. (1999) Formation of aerenchyma and the processes of plant ventilation in relation to soil flooding and submergence. *Plant Biology* 1, 274–287.

Jackson, M.B. & Drew, M. (1984). Effects of flooding on growth and metabolism of herbaceous plants. *In* Flooding and plant growth. T T Kozlowski (ed). pp. 47–128. Academic Press Inc., Orlando, Florida.

Jackson, M.B. & Hall, K.C. (1987). Early stomatal closure in flooded pea plants is mediated by abscisic acid in the absence of foliar water deficits. *Plant, Cell & Environment* 10, 121–130.

Jackson, M.B., Drew M.C. & Giffard S.C. (1981). Effects of applying ethylene to the root system of *Zea mays* L. on growth and nutrient concentration in relation to flooding. *Physiologia Plantarum* 52, 23–28.

Jackson, M.B., Saker, L.R., Crisp, C.M., Else, M.A. & Janowiak F. (2003). Ionic and pH signalling from roots to shoots of flooded tomato plants in relation to stomatal closure. *Plant and Soil* 253, 103–113.

Jackson, M.B., Fenning, T.M., Drew, M.C. & Saker, L.R. (1985). Stimulation of ethylene production and gas-space (aerenchyma) formation in adventitious roots of *Zea mays* L. by small partial pressures of oxygen. *Planta* 165, 486–492.

James, E.K. & Sprent, J.I. (1999). Development of N_2-fixing nodules on the wetland legume *Lotus uliginosus* exposed to conditions of flooding. *New Phytologist* 142, 219–231.

Justin, S.H.F.W. & Armstrong, W. (1987) The anatomical characteristics of roots and plant response to soil flooding. *New Phytologist* 106, 465–495.

Kotula, L., Ranathunge, K., Schreiber L. & Steudle, E. (2009). Functional and chemical comparison of apoplastic barriers to radial oxygen loss in roots of rice (*Oryza sativa* L.) grown in aerated or deoxygenated solution *Journal of Experimental Botany* 60, 2155–2167.

Kozlowski, T.T. & Pallardy, S.G. (1984). Effects of flooding on water, carbohydrate and mineral relations. *In* Flooding and plant growth. T T Kozlowski (ed). pp. 165–193. Academic Press Inc., Orlando, Florida.

Kozlowski, T.T. (1997). Responses of woody plants to flooding and salinity. Tree Physiology Monograph 1. http://www.heronpublishing.com/tp/monograph/kozlowski.pdf [Verified 28 July 2011].

Kulichikhin, K.Y., Greenway, H., Bryne, L. & Colmer T.D. (2009). Regulation of intracellular pH during anoxia in rice coleoptiles in acid and near neutral conditions. *Journal of Experimental Botany* 60, 2119–2128.

Laan, P., Tosserams, M., Blom, C.W.P.M. & Veen, B.W. (1990). Internal oxygen transport in *Rumex species* and its significance for respiration under hypoxic conditions. *Plant & Soil* 122, 39–46.

Lenssen, J.P.M., Van de Steeg, H.M. & de Kroon, H. (2004). Does disturbance favour weak competitors? Mechanisms of altered plant abundance after flooding. *Journal of Vegetation Science* 15, 305–314.

Liao, C.T. & Lin, C.H. (1994). Effect of flooding stress on photosynthetic activities of Momordica charantia. *Plant Physiology & Biochemistry* 32, 479–485

Liao, C.T. & Lin, C.H. (2001). Physiological adaptation of crop plants to flooding stress. *Proceeding of the National Science Council, Republic of China Part B* 25, 148–157.

Liu, J.H. & Reid, D.M. (1992). Auxin and ethylene–stimulated advantitious rooting in relation to tissue sensitivity to auxin and ethylene production in sunflower hypocotyls. *Journal of Experimental Botany* 43, 1191–1198.

Loreti, J. & Oesterheld, M. (1996). Intraspecific variation in the resistance to flooding and drought in populations of *Paspalum dilatatum* from different topographic positions. *Oecologia* 108, 279–284.

Lynn, D.E. & Waldren, S. (2003). Survival of *Ranunculus repens* L. (creeping buttercup) in an amphibious habitat. *Annals of Botany* 91, 75–84.

Malik, A.I., Colmer, T.D., Lambers, H. & Schortemeyer, M. (2001). Changes in physiological and morphological traits of roots and shoots of wheat in response to different depths of waterlogging. *Australian Journal of Plant Physiology* 28, 1121–1131.

Malik, A.I., Colmer, T.D., Lambers, H., Setter, T.L. & Schortemeyer, M. (2002). Short–term waterlogging has long–term effects on the growth and physiology of wheat. *New Phytologist* 153, 225–236.

Malik, A.I., Islam, A.K.M.R. & Colmer, T.D. (2011). Transfer of the barrier to radial oxygen loss in roots of *Hordeum marinum* to wheat (*Triticum aestivum*): evaluation of four *H. marinum*–wheat amphiploids. *New Phytologist* 190, 499–508.

Mancuso, S. & Shabala, S. (2010). Waterlogging signalling and tolerance in plants. Pp 294. Springer–Verlag Berlin, Heidelberg.

Manzur, M.E., Grimoldi, A.A., Insausti, P. & Striker G.G. (2009). Escape from water or remain quiescent? *Lotus tenuis* changes its strategy depending on depth of submergence. *Annals of Botany* 104, 1163–1169.

McDonald, M.P. & Visser, E.J.W. (2003). A study of the interaction between auxin and ethylene in wildtype and transgenic ethylene insensitive tobacco during adventitious root formation induced by stagnant root zone conditions. *Plant Biology* 5, 550–556.

McDonald, M.P., Galwey, N.W. & Colmer, T.D. (2002). Similarity and diversity in adventitious root anatomy as related to root aeration among a range of wet– and dry–land grass species. *Plant, Cell & Environment* 25, 441–451.

Mielke, M.S., Almeida, A–AF., Gomes, F.P., Aguilar, M.A.G. & Mangabeira, P.A.O. (2003). Leaf gas exchange, chlorophyll fluorescence and growth responses of *Genipa americana* seedlings to soil flooding. *Environmental & Experimental Botany* 50, 221–231.

Mollard, F.P.O., Striker, G.G., Ploschuk, E.L. & Insausti, P. (2010). Subtle topographical differences along a floodplain promote different plant strategies among *Paspalum dilatatum* subspecies and populations. *Austral Ecology* 35, 189–196.

Mollard, F.P.O., Striker, G.G., Ploschuk, E.L., Vega, A.S. & Insausti, P. (2008). Flooding tolerance of *Paspalum dilatatum* (Poaceae: Paniceae) from upland and lowland positions in a natural grassland. *Flora* 203, 548–556.

Mommer, L. & Visser, E.J.W. (2005). Underwater photosynthesis in flooded terrestrial plants: a matter of leaf plasticity. *Annals of Botany* 96, 581–589.

Mommer, L., Pedersen, O. & Visser, E.J.W. (2004). Acclimation of a terrestrial plant to submergence facilitates gas exchange under water. *Plant, Cell & Environment* 27, 1281–1287.

Murchie, E.H. & Niyogi, K.K. (2011). Manipulation of photoprotection to improve plant photosynthesis. *Plant Physiology* 155, 86–92.

Naidoo, G. & Mundree, S.G. (1993). Relationship between morphological and physiological responses to waterlogging and salinity in *Sporobolus virginicus* (L.) Kunth. *Oecologia* 93, 360–366.

Noctor, G. & Foyer, C.H. (1998). Ascorbate and glutathione: Keeping active oxygen under control. *Annual Review of Plant Physiology and Plant Molecular Biology* 49, 249–279.

Parelle, J., Roudaut, J-P. & Ducrey, M. (2006). Light acclimation and photosynthetic response of beech (*Fagus sylvatica* L.) saplings under artificial shading or natural Mediterranean conditions. *Annals of Forest Science* 63, 257–266.

Parent, C., Capelli, N. & Dat, J. (2008). Reactive oxygen species, stress and cell death in plants. *Comptes Rendus – Biologies* 331, 255–261.

Parolin, P. (2009). Submerged in darkness: adaptations to prolonged submergence by woody species of the Amazonian floodplains. *Annals of Botany* 103, 359–376.

Pedersen, O., Binzer, T. & Borum, J. (2004). Sulfide intrusion in eelgrass (*Zostera marina* L.). *Plant, Cell & Environment* 27, 595–602.

Pereira, J.S. & Kozlowski, T.T. (1977). Variations among woody angiosperms in response to flooding. *Physiologia Plantarum* 41, 184–192.

Pezeshki, S.R. (1994). Responses of baldcypress (*Taxodium distichum*) seedlings to hypoxia: Leaf protein content, ribulose-1,5-bisphosphate carboxylase/oxygenase activity and photosynthesis. *Photosynthetica* 30, 59–68.

Pezeshki, S.R. (1996). Responses of three bottomland species with different flood tolerance capabilities to various flooding regimes. *Wetlands Ecology & Management* 4, 245–256.

Pezeshki, S.R. (2001). Wetland plant responses to soil flooding. *Environmental &Experimental Botany* 46, 299–312.

Pezeshki, S.R. & DeLaune, R.D. (1998). Responses of seedlings of selected woody species to soil oxidation-reduction conditions. *Environmental & Experimental Botany* 40, 123–133.

Pierik, R., van Aken, J.M. & Voesenek L.A.C.J. (2009). Is elongation–induced leaf emergence beneficial for submerged *Rumex* species? *Annals of Botany* 103, 353–357.

Ponnamperuma, F.N. (1972). Chemistry of submerged soils. *Advances in Agronomy* 24, 29–95.

Ponnamperuma, F.N. (1984). Effects of flooding on soils. *In* Flooding and Plant Growth. T.T. Kozlowski (ed). pp 9–45. Academic Press, Orlando, Florida.

Postma, J.A. & Lynch, J.P. (2011). Root cortical aerenchyma enhances the growth of maize on soils with suboptimal availability of nitrogen, phosphorus, and potassium. *Plant Physiology* 156, 1190–1201.

Ram, P.C., Singh, B.B., Singh, A.K., Ram, P., Singh, P.N,. Singh, H.P., Boamfa, I., Harren, F., Santosa, E., Jackson, M.B., Setter, T.L., Reuss, J., Wade, L.J., Pal Singh, V. & Singh, R.K. (2002). Submergence tolerance in rainfed lowland rice: Physiological basis and prospects for cultivar improvement through marker–aided breeding. *Field Crops Research* 76, 131–152.

Ranathunge, K., Lin, J., Steudle, E. & Schreiber, L. (2011). Stagnant deoxygenated growth enhances root suberization and lignifications, but differentially affects water and NaCl permeabilities in rice (*Oryza sativa* L.) roots. *Plant, Cell & Environment* 34, 1223–1240.

Sachs, M. & Vartapetian, B. (2007). Plant anaerobic stress I. Metabolic adaptation to oxygen deficiency. *Plant Stress* 1, 123–135.

Sand–Jensen, K. (1989). Environmental variables and their effect on photosynthesis of aquatic plant communities. *Aquatic Botany* 34, 5–25.

Schussler, E.E. & Longstreth, D.J. (2000). Changes in cell structure during the formation of root aerenchyma in *Sagittaria lancifolia* (Alismataceae). *American Journal of Botany* 87, 12–19.

Schlüter, U. & Crawford, R.M.M. (2001). Long–term anoxia tolerance in leaves of *Acorus calamus* L. and *Iris pseudacorus* L. *Journal of Experimental Botany* 52, 2213–2225.

Seago, J.L., Marsh, L.C., Stevens, K.J., Soukup, A., Vortubová, O. & Enstone, D.E. (2005). A re–examination of the root cortex in wetland flowering plants with respect to acrenchyma. *Annals of Botany* 96, 565–579.

Setter, T.L. & Laureles, E.V. (1996). The beneficial effect of reduced elongation growth on submergence tolerance of rice. *Journal of Experimental Botany* 47, 1551–1559.

Shimamura, S., Mochizuki, T., Nada, Y. & Fukuyama M. (2003). Formation and function of secondary aerenchyma in hypocotyl, roots and nodules of soybean (*Glycine max*) under flooded conditions. *Plant & Soil* 251, 351–359.

Shimamura, S., Yoshida, S. & Mochizuki, T. (2007). Cortical aerenchyma formation in hypocotyl and adventitious roots of *Luffa cylindrica* subjected to soil flooding. *Annals of Botany* 100, 1431–1439.

Shimamura, S., Yamamoto, R., Nakamura, T., Shimada, S. & Komatsu, S. (2010). Stem hypertrophic lenticels and secondary aerenchyma enable oxygen transport to roots of soybean in flooded soil. *Annals of Botany* 106, 277–284.

Shiono, K., Ogawa, S., Yamazaki, S., Isoda, H., Fujimura, T., Nakazono, M. & Colmer. T.D. (2011). Contrasting dynamics of radial O_2–loss barrier induction and aerenchyma formation in rice roots of two lengths. *Annals of Botany* 107, 89–99.

Smirnoff, N. & Crawford, R.M.M. (1983) Variation in the structure and response to flooding of root aerenchyma in some wetland plants. *Annals of Botany* 51, 237–249.

Soriano, A. (1991). Río de la Plata Grasslands. In: Ecosystems of the world 8A. Natural grasslands. Introduction and Western Hemisphere, pp. 367–407 (R T Coupland Ed). Elsevier, Amsterdam, The Netherlands.

Soukup, A., Votrubova, O. & Cizkova, H. (2002). Development of anatomical structure of roots of *Phragmites australis*. *New Phytologist* 153, 277–287.

Stevens, K.J., Peterson, R.L. & Stephenson, G.R. (1997). Morphological and anatomical responses of Lythrum salicaria L. (purple loosestrife) to an imposed water gradient. *International Journal of Plant Sciences* 158, 172–183.

Striker, G.G. (2008). Visiting the methodological aspects of flooding experiments: Quantitative evidence from agricultural and ecophysiological studies. *Journal of Agronomy & Crop Science* 194, 249–255.

Striker, G.G., Insausti, P. & Grimoldi, A.A. (2007b). Effects of flooding at early summer on plant water relations of *Lotus tenuis*. *Lotus Newsletter* 37, 1–7.

Striker, G.G., Insausti, P. & Grimoldi, A.A. (2008). Flooding effects on plant recovery from defoliation in the grass *Paspalum dilatatum* and the legume *Lotus tenuis*. *Annals of Botany* 102, 247–254.

Striker, G.G., Manzur, M.E. & Grimoldi, A.A. (2011a). Increasing defoliation frequency constrains regrowth of *Lotus tenuis* under flooding. The role of crown reserves. *Plant & Soil* 343, 261–272.

Striker, G.G., Insausti, P., Grimoldi, A.A. & León, R.J.C. (2006). Root strength and trampling tolerance in the grass *Paspalum dilatatum* and the dicot *Lotus glaber* in flooded soil. *Functional Ecology* 20, 4–10.

Striker, G.G., Insausti, P., Grimoldi, A.A. & Vega, A.S. (2007a). Trade–off between root porosity and mechanical strength in species with different types of aerenchyma. *Plant, Cell & Environment* 30, 580–589.

Striker, G.G., Izaguirre, R.F., Manzur, M.E. & Grimoldi, A.A. (2011b). Different strategies of *Lotus japonicus*, *L. corniculatus* and *L. tenuis* to deal with complete submergence at seedling stage. *Plant Biology* (doi:10.1111/j.1438–8677.2011.00493.x)

Striker, G.G., Insausti, P., Grimoldi, A.A., Ploschuk, E.L. & Vasellati, V. (2005). Physiological and anatomical basis of differential tolerance to soil flooding of *Lotus corniculatus* L. and *Lotus glaber* Mill. *Plant & Soil* 276, 301–311.

Teakle, N.L., Armstrong, J., Barrett–Lennard, E.G & Colmer, T.D. (2011). Aerenchymatous phellem in hypocotyl and roots enables O_2 transport in *Melilotus siculus*. *New Phytologist* 190, 340–350.

Titarenko, T.Y. (2000). Test parameters of revealing the degree of fruit plants tolerance to the root hypoxia caused flooding of soil. *Plant Physiology & Biochemistry* 38, 115.

Tournaire–Roux, C., Sutka, M., Javot, H., Gout, E., Gerbeau, P., Luu, D.T., Richard Bligny, R. & Maurel, C. (2003). Cytosolic pH regulates root water transport during anoxic stress through gating of aquaporins. *Nature* 425, 393–397.

Vartapetian, B.B. & Jackson, M. (1997). Plant adaptations to anaerobic stress. *Annals of Botany* 79, 3–20.

Vashisht, D., Hesselink, A., Pierik, R., Ammerlaan, J.M.H., Bailey-Serres, J., Visser, E.J.W., Pedersen, O., van Zanten, M., Vreugdenhil, D., Jamar, D.C.L., Voesenek L.A.C.J. & Sasidharan, R. (2011). Natural variation of submergence tolerance among *Arabidopsis thaliana* accessions. *New Phytologist* 190, 299–310.

Verdoucq, L., Grondin, A. & Maurel, C. (2008). Structure–function analyses of plant aquaporin AtPIP2;1 gating by divalent cations and protons. *The Biochemical Journal* 415, 409–416.

Visser, E.J.W. & Bögemann, G.M. (2006). Aerenchyma formation in the wetland plant *Juncus effusus* is independent of ethylene. *New Phytologist* 171, 305–314.

Visser, E.J.W. & Voesenek, L.A.C.J. (2004). Acclimation to soil flooding – sensing and signal-transduction. *Plant & Soil* 254,197–214.

Visser, E.J.W., Cohen, J.D., Barendse, G.W.M., Blom, C.W.P.M. & Voesenek, L.A.C.J. (1996). An ethylene–mediated increase in sensitivity to auxin induces adventitious root formation in flooded *Rumex palustris* Sm. *Plant Physiology* 112, 1687–1692.

Visser, E.J.W., Colmer, T.D., Blom, C.W.P.M & Voesenek, L.A.C.J. (2000). Changes in growth, porosity, and radial oxygen loss from adventitious roots of selected mono- and dicotyledonous wetland species with contrasting types of aerenchyma. *Plant, Cell & Environment* 23, 1237–1245.

Visser, E.J.W., Heijink, C.J., Van Hout, K.J.G.M., Voesenek, L.A.C.J., Barendse, G.W.M. & Blom, C.W.P.M. (1995). Regulatory role of auxin in adventitious root formation in two species of *Rumex*, differing in their sensitivity to waterlogging. *Physiologia Plantarum* 93, 116–122.

Voesenek, L.A.C.J., Harren, F.J., Bögemann, G.M., Blom, C.W.P.M. & Reuss, J. (1990). Ethylene production and petiole growth in *Rumex* plants induced by soil waterlogging: the application of a continuous flow system and a laser driven intracavity photoacoustic detection system. *Plant Physiology* 94, 1071–1077.

Voesenek, L.A.C.J., Banga, M., Thier, R.H., Mudde, C.M., Harren, F.J.M., Barendse, G.W.M. & Blom, C.W.P.M. (1993). Submergence–induced ethylene synthesis, entrapment, and growth in two plant species with contrasting flooding resistances. *Plant Physiology* 103, 783–791.

Voesenek, L.A.C.J., Rijnders, J., Peeters,A.J.M.,Van de Steeg H.M.V. & De Kroon, H. (2004). Plant hormones regulate fast shoot elongation under water: from genes to communities. *Ecology* 85, 16–27.

Voesenek, L.A.C.J., Colmer, T.D., Pierik, R., Millenaar, F.F. & Peeters, A.J.M. (2006). How plants cope with complete submergence. *New Phytologist* 170, 213–226.

Vriezen, W.H., De Graaf, B., Mariani, C. & Voesenek, L.A.C.J. (2000). Submergence induces expansin gene expression in flooding–tolerant *Rumex palustris* and not in flooding intolerant *R. acetosa*. *Planta* 210, 956–963.

Vu, J.C.V. & Yelenosky, G. (1991). Photosynthetic responses of citrus trees to soil flooding. *Physiologia Plantarum* 91, 7–14.

Wample, R.L. & Davis, R.W. (1983). Effect of flooding on starch accumulation in chloroplasts of sunflower (*Helianthus annuus* L.). *Plant Physiology* 73, 195–198.

Wegner, L.H. (2010). Oxygen transport in waterlogged plants. *In* Waterlogging Signalling and Tolerance in Plants. S Mancuso & S Shabala (eds). pp. 3–22. Springer-Verlag Berlin, Heidelberg.

Xu, K., Xu, X., Fukao, T., Canlas, P., Marghirang-Rodriguez, R., Heuer, S., Ismail, A.M., Bailey-Serres, J., Ronald, P.C. & Mackill, D.J. (2006). *Sub1A* is an ethylene-response-factor-like gene that confers submergence tolerance to rice. *Nature* 442, 705–708.

Yordanova, R.Y. & Popova, L.P. (2001). Photosynthetic response of barley plants to soil flooding. *Photosynthetica* 39, 515–520.

Yordanova, R., Christov, K. & Popova, L. (2004). Antioxidative enzymes in barley plants subjected to soil flooding. *Environmental & Experimental Botany* 51, 93–101.

3

Review on Some Emerging Endpoints of Chromium (VI) and Lead Phytotoxicity

Conceição Santos and Eleazar Rodriguez
Laboratory of Biotechnology and Cytometry, Department of Biology & CESAM,
University Aveiro, Aveiro
Portugal

1. Introduction

Metals occur naturally in the environment as constituents of the Earth's crust and they tend to accumulate and persist due to their stability and mainly because they cannot be degraded or destroyed. However, and despite that in some cases (e.g. mercury) high levels occur naturally, for most situations, anthropic activities are among the primary causes for metal pollution. Examples of important sources of metal contamination come from industrial applications, mining, smelters, combustion by-products and fuel. From these sources, contaminants can enter the ecosystem as airborne particles, wastewaters and sludge, polluting not only sites near the source but locations thousands of kilometers apart. Studies like the ones of Murozumi et al. (1969), Hong et al. (1994) or McConnell and Edwards (2008) demonstrated the extension and persistence of metals in the environment. These studies also showed that contamination of the environment with these pollutants started way before the industrial revolution with evidence of pollution originating from Roman mining and smelters in 500 B.C. (Nriagu, 1996).

Due to the above reasons and to their toxicity to human health and environment, metal toxicity has become an increasing target of studies in humans, animals and plants. Of what is generally conceived, toxicity originates through a very complex pattern of metal interactions with cellular macromolecules, metabolic and signal transduction pathways and genetic processes (Beyersmann and Hartwig, 2008). Among the different models available to study metal toxicity, plants present some unique features that make them interesting subjects. Firstly, much of human diet depend directly from plants products like fruits and vegetables or indirectly as fodder given to livestock. Secondly, by lacking the ability to escape from contaminated sites, plants evolved mechanisms to handle exposure to toxicants, from the amount that is taken from the surroundings, to strategies of sequestration and inactivation in sub cellular compartments or even to the ability of tolerating putative deleterious effects of metals.

Regarding the amount of pollutant accumulated, three categories of plants were proposed by Baker (1981): *(1) excluders*: those that grow in metal-contaminated soil and maintain the shoot concentration at low level up to a critical soil value above which relatively unrestricted root-to-shoot transport results; *(2) accumulators*: those that concentrate metals in the aerial part; (3) *indicators*: where uptake and transport of metals to the shoot are regulated so that internal concentration reflects external levels, at least until toxicity occurs.

The toxicity of metals, and of their compounds, largely depends on their bioavailability, i.e. the mechanisms of uptake through the cell's membrane, intracellular distribution and binding to cellular macromolecules (Beyersmann and Hartwig, 2008). Although the relative toxicity of different metals to plants can vary with plant genotype and experimental conditions, most act through one of the following: changes in the permeability of the cell's membrane; reactions of sulphydryl (–SH) groups with cations; affinity for reacting with phosphate groups and active groups of ADP or ATP; replacement of essential ions and oxidative stress (Patra et al., 2004). Through these, some of the most common, and often unspecific symptoms, of metals phytotoxicity are: growth inhibition, nutrient imbalance, disturbances in the ion and water regime (e.g. Gyuricza et al., 2010), photosynthetic impairment (e.g. Hattab et al., 2009b) and genotoxicity (e.g. Monteiro et al., 2010).

Most of the metals of greater environmental concern have been currently included in the classical and ill-defined group of "heavy metals". This is an unclear term for a group of elements that present metallic proprieties and normally include transition metals, some metalloids, lanthanides, and actinides. Some years ago, this term has been considered meaningless and misleading by the IUPAC due to the contradictory definitions and its lack of a coherent scientific basis (Duffus, 2002). It has been since then progressively abandoned by the scientific community, but still remains widespread in many reports, mostly reporting to any metallic element with relatively high density and which is toxic in low concentrations. Among the elements referred to as "heavy metals", 13 have been considered by the European Union to be of the highest concern: arsenic (As), cadmium (Cd), cobalt (Co), chromium (Cr), copper (Cu), mercury (Hg), manganese (Mn), nickel (Ni), lead (Pb), tin (Sn) and thallium (Ti). From these, some have been target of many investigations (e.g. Cd) while for others, the level of knowledge about the mechanism of toxicity are highly unsatisfactory (e.g. Cr). Interestingly, despite it being one of the first metals with known reports of human poisoning, not enough research have been undertaken to clarify Pb's mechanism of toxicity and even some conflicting data have been reported (García-Lestón et al., 2010).

In this review we´ll discuss some of the most relevant and updated data on Cr and Pb toxicity in plant cells, and explore some of the emerging techniques to diagnose cyto and genotoxicity.

2. Chromium: The element

Chromium was discovered in 1797 as part of the mineral crocoite, used as pigment due to its intense coloration. As a matter of fact, the name chromium is derived from the Greek word "χρώμα" (chroma- color) due to that propriety of the element. Chromium is the 21st most abundant element in Earth's crust with an average concentration of 100 ppm, ranging in soil between 1 mg/kg and 3000 mg/kg; in sea water from 5 µg/L to 800 µg/L and in rivers and lakes between 26 µg/L and 5.2 mg/L. Normally, Cr is mined from chromate but native deposits are not unheard off. One of the most interesting characteristics of this metal is its hardness and high resistance to corrosion and discoloration. The importance of these proprieties resulted among others in the usage of this metal in the development of stainless steel, which together with chrome plating and leather tanning, are the most important applications of this element and the main sources of Cr pollution of the environment. Chromium is highly soluble under oxidizing conditions and forms, exhibiting a wide range of possible oxidation states (from -2 to +6), being that +3 [Cr(III)] and +6 [Cr(VI)] are the most stable forms. Under reducing conditions, Cr(VI) converts to Cr(III) that is insoluble, but this form is strongly absorbed onto the surface of soil particles.

3. Chromium: Uptake and assimilation by plants

Chromium is a common contaminant of surface waters and ground waters because of its occurrence in nature, as well as anthropic sources (Babula et al., 2008). Cr(III) and Cr(VI), being the most stable are also the important in terms of environmental contamination. The most important sources of Cr(III) are fugitive emissions from road dust and industrial cooling towers; also, Cr(VI) compounds are still used in the manufacture of pigments, in metal-finishing and chromium-plating, in stainless steel production, in hide tanning, as corrosion inhibitors, and in wood preservation (Shtiza et al. 2008).

Very few studies have attempted to elucidate the transport mechanisms of Cr in plants, but factors like oxidative Cr state or its concentration in substrate play important roles (Babula et al., 2008). Of what is known, due to its higher solubility and thus, bioavailability, Cr(VI) is more toxic at lower concentrations than Cr(III), which tend to form stable complexes in soils (Lopez-Luna et al., 2009). Also, the pathway of Cr(VI) transport is thought to be an active mechanism involving carriers of essential anions such as sulfate (Cervantes et al., 2001). Fe, S and P are known also to compete with Cr for carrier binding (Wallace et al., 1976). Also Cr absorption and translocation have been show to be modified by soil pH, organic matter content and chelating agents, among others (Han et al., 2004).

Studies performed to elucidate the uptake mechanism of Cr have demonstrated that only Cr(VI) is detected in plant tissues. However some plants (such as soybean and garlic) have the capacity to reduce Cr(VI) to unstable intermediate like Cr(V) and Cr(IV), or eventually to the more stable form, Cr(III); this represents the detoxification pathway of Cr(VI) (Babula et al., 2008). As this mechanism of detoxification is performed readily in the roots and as Cr is immobilized in the vacuoles of the root cells, the amount of Cr translocated to the aerial portion of the plants is very little (Shanker et al., 2005).

4. Chromium: Phytotoxicity

4.1 General effects

The effects of Cr in some of the classical endpoints of heavy metal genotoxicity have received some attention by fellow researchers. Seed germination and plants growth are two of the parameters that have been studied thoroughly. Results indicate that Cr provokes growth inhibition of roots in species like *Salix viminalis* (Prasad et al., 2001), *Caesalpinia pulcherrima* (Iqbal et al., 2001), wheat (Chen et al., 2001) and mung bean (Samantaray et al., 1998).

Shanker et al. (2005) hypothesized that root growth inhibition due to Cr toxicity could be due to inhibition of root cell division/root elongation or to the extension of cell cycle. Aerial part growth (measured by effects on shoot length and on reduction of leaf number and area) has also been proven to be negatively affected by Cr in species like rice (Singh et al., 2006) wheat, oat and sorghum (Lopez-Luna et al., 2009). Justifications to these facts were proposed by Shanker et al. (2005) by stating that root growth inhibition, as well as its consequent and/or causal nutrient imbalance, could be behind low shoot development. In fact, chromium, due to its structural similarity with some essential elements, can affect mineral nutrition of plants in a complex way (Shanker et al., 2005) and there has been innumerous considerations regarding this issue, especially in crop species.

It has been demonstrated that very low concentrations of Cr (0.05–1 mg /L) promoted growth, and increased nitrogen fixation and yield in leguminosae (e.g. Hewitt, 1953). At higher concentrations and just giving a couple of examples, authors have successfully proven that this metal reduces the uptake of the essential elements Fe, K, Mg, Mn, P and Ca in *Salsola kali* (Gardea-Torresdey et al., 2005) and K, Mg, P, Fe and Mn in roots of soybean (Turner and Rust, 1971). The justification of nutrient imbalance has been pointed to competitive binding to common carriers by Cr(VI), to inhibition of the activity of plasma membrane's H^+-ATPase and to reduced root growth and impaired penetration of the roots into the soil due to Cr toxicity (Shanker et al., 2005).

4.2 Photosynthesis

Like other metals, Cr can affect photosynthesis severally and in many different steps, which can ultimately translate in loss of productivity and death. Shanker et al. (2005), in a review about Cr phytotoxicity, discussed that while Cr toxicity at the photosynthetic level was well documented in trees and higher plants, the exact target and mechanisms affected by this metal were poorly understood.

Cr(VI) can easily cross biological membranes and has high oxidizing capacity, generating reactive oxygen species (ROS) which might induce oxidative stress (Pandey et al., 2009). ROS are generated in normal metabolic processes like respiration and photosynthesis, being chloroplasts one of the main sites of reactive oxygen production and detoxification (Mittler, 2002). However, because the chloroplast has high amounts and complex systems of membranes rich in polyunsaturated fatty acids, this organelle might also be a target for peroxidation (Hattab et al., 2009b) and one of the ways by which photosynthesis is affected. A common parameter affected by Cr is the amount of photosynthetic pigments, which tends to decrease when plants or algae are exposed to high doses of this metal (Rodriguez et al., 2011, Subrahmanyam, 2008, Vernay et al., 2007). The results obtained by Juarez et al. (2008) using algae, demonstrated that, ROS caused structural damage to the pigment-protein complexes located in the thylakoid membrane (e.g. the destabilization and degradation of the proteins of the peripheral part of antenna complex), followed by the pheophytinization of the chlorophylls (substitution of Mg^{2+} by H^+ ions), and destruction of the thylakoid's membranes. It has also been demonstrated that Cr affects, and might even inhibit, pigment biosynthesis, among others, by degrading δ-aminolaevulinic acid dehydratase (Vajpayee et al., 1999), an essential enzyme in chlorophyll biosynthesis. Vernay et al. (2007) also presented evidence that this metal probably competed with Fe and Mg for assimilation and transport to the leaves and therefore affected different steps of pigment biosynthesis.

Another endpoint of Cr phytotoxicity is *Chl a* fluorescence; however, it was demonstrated that, within some of the common biomarkers of *Chl a* fluorescence, most parameters evaluated are somewhat resistant to Cr toxicity (namely the F_v/F_m). On the other hand, the ones related to the fluorescence emission status of light adapted-plants have been shown to be highly affected by this metal (Subrahmanyam, 2008, Vernay et al., 2007). Several hypotheses explaining these results have been proposed, e.g., structural alterations in the pigment–protein complexes of PSII or impairment in energy transfer from antennae to reaction centers (like a diversion of electrons from the electron-donating side of PS I to Cr(VI) are the most endorsed (Shanker et al., 2005). Recently, Henriques (2010) implied that Cr(VI) might not be directly responsible for the damage to the chloroplast, as the valence state of Cr depends of the local pH and redox values. For instance, in irradiated chloroplasts, the previously mentioned conditions would favor the less toxic Cr(III) form over the highly toxic Cr(VI). Appenroth et al. (2000) demonstrated that Cr damaged the water oxidizing centers (WOC) associated to PSII and Henriques (2010) proposed that this could be explained by the reduction of the Ca and Mn availability, caused by Cr, which are fundamental in the structure and functioning of the WOC.

Besides the photochemical process, Cr is also known to cause distress in the biochemical aspects of photosynthesis. Vernay et al. (2007, 2008) discussed that despite that loss of biomass and wilting were common symptoms of Cr exposure, little was known about Cr effect on water status and gas exchange. Subrahmanyam (2008) also commented that it was unclear if Cr-induced inhibition of the photosynthetic process was also due (among others previously mentioned factors) to Cr-induced interference with the Calvin cycle's enzymes. In those reports, the authors proved that Cr consistently affected parameters like E (transpiration rate), g_s (stomatal conductance), A (photosynthetic rate) and C_i (substomatal CO_2 concentration). One of the main conclusions of those articles was that even though the decrease in g_s seemed to be responsible for the variation in water regulation status, the increase in Ci induced by Cr accumulation clears g_s as the responsible for the decrease in A. This also indicates as hinted by Subrahmanyam (2008) and by Vernay et al. (2007) that the reduction in A might lay in the functional status of the Calvin cycle enzymes. Unfortunately, the availability of data regarding Cr putative effects on the enzymes of the Calvin cycle is far less than what exists for other parameters.

The recent works of Dhir et al. (2009) and Bah et al. (2010) provided one of the first insights to Cr-induced effects at the Calvin cycle enzymes. Dhir et al. (2009) found a significant decrease in ribulose-bisphosphate carboxylase oxygenase (RuBisCO) activity induced by exposure to wastewaters (rich in Cr) from an electroplating unit and suggested that this results could be explained by: a substitution of Mg^{2+} in the active site of RuBisCO subunits by metal ions; decline in RuBisCO content as a result of oxidative damage; a shift in the enzyme's activity from carboxilation to oxygenation. On the other hand, Bah et al. (2010) performed a proteomic analysis of *Typha angustifolia*'s leaves exposed to metals and found that exposure to Cr induced the expression of ATP synthase, RuBisCO small subunit and coproporphyrinogen III oxidase. The authors then explained that their data were an evidence of a protective mechanism against metal toxicity at the photosynthetic level, which might be responsible for the metal tolerance displayed by *T. angustifolia*. Furthermore, the authors also suggested that the increased expression of ATP synthase was indicative of the high energetic requirements needed to cope with metal toxicity.

Recently we compared the Cr(VI) phytotoxicity using some photosynthetic endpoints in pea leaves exposed to this metal (up to 2000 mg / L) (Rodriguez, 2011). Our group demonstrated that Cr(VI) was more aggressive to the gas exchange, biochemical and chloroplastidial morphology markers than to those related to the photochemical apparatus. However, exposure to higher Cr(VI) dosages induced significant negative effects on the photochemical apparatus, proving that despite having some degree of resistance to metal toxicity it can still be damaged by Cr(VI) (Rodriguez, 2011). In these analyses metal toxicity in photosynthesis, flow cytometry (FCM) was used in complement to the classical tools. Few reports have up to moment tried to apply FCM's potential to study chloroplast and there are even less reports focused on evaluating the effects of hazardous substances in these organelles. In that assay our group compared the information provided by FCM vs PAM fluorometry and pigment content was also carried, in order to assess if chloroplast auto-fluorescence emission, as measured by FCM, related to those classical techniques.

FCM is gaining importance in toxicological assays of the photosynthetic machinery. We demonstrated FCM reliability and its endowment of complementary data to conventional techniques as PAM in a recent exhaustive study with paraquat (Rodriguez et al., 2011c; Figure 1). In that study, FCM and PAM fluorometry presented a strong positive correlation value, even though FCM measurements were performed on isolated chloroplasts while for

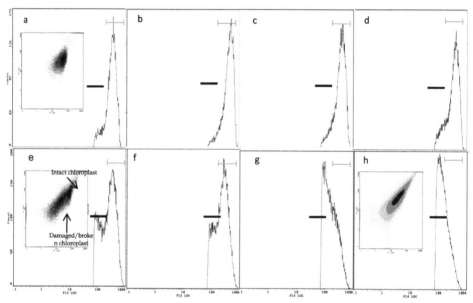

Fig. 1. Validation of FCM as measuring chloroplast fluorescence after stress: Histograms of relative fluorescence intensity (FL) of chloroplast isolated form plants exposed to a contaminant (paraquat). a) control, b) 3h of exposure, c) 6h of exposure, d) 9h of exposure, e) 12h of exposure, f) 15h exposure, g) 18h exposure, h) 24h exposure (Due to its similarity, the histogram for 21h of exposure was omitted). In each histogram 2 regions are marked: the ⊢——⊣ region that defines population B (chloroplast with higher integrity, with higher fluorescence intensity and that shows a tendency to desapear with the increase of stress throughout the time); the ━━ region that has lower fluorescence intensity and appears to dominate with the stress. Inserted in histograms a), e) and h), are the respective cytograms of FS (volume) vs SS (granularity) of the isolated chloroplasts in a logarithmic scale. (Adapted from Rodriguez et al 2011c; Rodriguez 2011).

PAM fluorometry, intact leaves (i.e. still part of the plants) were used (Rodriguez, 2011; Rodriguez et al., 2011c). Also FCM was used in *Chlorell vulgaris* cultures: volume, granularity and algal autofluorescence intensity (FL) were determined by FCM, and it was demonstrated that algal density was the most affected parameter measured, while cell volume and, less, granularity were affected in a similar manner (Rodriguez et al., 2011 a).

Another unexplored endpoint of Cr-induced stress at the metabolic level is the variation in the amount of soluble sugars and starch accumulated in leaves. Besides being the fuel for carbon and energy metabolism, sugars also play a pivotal role as signaling molecules (Rolland et al., 2006). Therefore, the quantification of the sugar levels in the leaves could provide information of paramount importance in the characterization and understanding of Cr-induced phytotoxicity.

Despite the appalling lack of data, reports like the ones presented by Tiwari et al. (2009) and Prado et al. (2010) offer some insight into the effects of Cr at this level. Tiwari and co-workers (2009) found that exposure to increasing concentration of Cr caused a decrease in the amount of non-reducing sugars while the inverse was observed for reducing sugars. Prado et al. (2010) on the other hand observed that Cr exposure caused the levels of sucrose (transport sugar) to increase while the concentration of glucose decreased.

4.3 Genotoxicity

In animals and yeast, Cr (VI) has been extensively studied and shown to be highly toxic, inducing cell cycle arrest and causing carcinogenic effects (e.g. O'Brien et al., 2002, Salnikow and Zhitkovich, 2007, Zhang et al., 2001). Despite of the critical importance of Cr toxicity, we are still far away of having in plants, the same level of understanding that exists in other eukaryotes about the mechanisms of Cr genotoxicity. What serves as bases for understanding Cr genotoxicity in plants is what is known in other organisms; Cr is a special case as unlike other metals, when inside the cell, Cr interacts primarily and directly with DNA, forming DNA-protein and DNA-DNA cross links, making this element a highly mutagenic and carcinogenic toxicant. By this, while other metals are considered weakly mutagenic, mostly acting through the inhibition of DNA repair machinery, Cr acts directly on DNA causing genotoxicity directly.

Cr can also form complexes which can react with hydrogen peroxide and generate significant amounts of hydroxyl radicals that may directly trigger DNA alterations and other effects (Shi and Dalal 1990a,b). The mechanism of Cr(VI) detoxification by reductases creates unstable forms of Cr that are known to create ROS which are one of the most common causes of DNA degradation. It has been shown that Cr(V) reacted with isolated DNA to produce 8-hydroxydeoxyguanosine, whereas Cr(VI) performed this reaction only in the presence of the reductant glutathione (Faux et al. 1992).

In cultured mammalian cells, Cr(VI) induced superoxide and nitric oxide production (Hassoun and Stohs 1995), whereas treatment of cells with Cr(VI) in the presence of glutathione reductase generated hydroxyl radicals. This ROS, besides degrading DNA, can also affect Mitogenic-Activated Protein Kinsases (MAPK), which cause the deregulation of cell proliferation (tumor inducing effect), thus causing mutagenicity through an indirect path, besides the aforementioned direct interaction with DNA (Beyersmann and Hartwig, 2008).

Cr genotoxicity studies in plants are summarized in Table 1. Most of the researches performed in plants have demonstrated that Cr generates chromosomal aberration and micronuclei formation, which is understandable, as both of these are commonly used as genotoxicity endpoints in ecotoxicological assay. As it can be seen in table 1, there is also evidence of Cr related DNA degradation (Comet assay) and point-mutation (AFLP), and thus it is very likely that more research will confirm that at least part of what is known in animals can also be observed in plants.

Species	Reference	Dose	Technique	Effects
V. faba	(Chandra et al., 2004)	Tannery solid waste	Cytogenetic	Chromosomal and mitotic aberration
B. napus	(Labra et al., 2004)	$K_2Cr_2O_7$ (10 to 200 mg/L)	AFLP, SAMPL, DNA methylation analysis	Methylation changes, Mutation
A. thaliana	(Labra et al., 2003)	$K_2Cr_2O_7$ (2, 4 and 6 mg/L)	AFLP	DNA mutation
C. sativa	(Citterio et al., 2003)	$K_2Cr_2O_7$ (25μg/g and 50 μg/g soil)	FCM	ND
T. repens	(Citterio et al., 2002)	Contaminated soils from a steelworks- Up to 4810 mg/kg soil	AFLP, FCM	Mutation, DNA decrease
V. faba	(Wang, 1999)	Cr (contaminated soils)	MN	Dose-related increase of MN
A. cepa	(Matsumoto et al., 2006)	Tannery effluent	Cytogenetic	Chromossomal aberration
V. faba	(Koppen and Verschaeve, 1996)	$K_2Cr_2O_7$ (Up to 10^{-3}M)	COMET assay	Increase in %Tail DNA, Tail moment and Tail length
Tradescancia sp. V. faba	(Knasmuller et al., 1998)	$CrCl_3$, CrO_3 (from 0.75 to 10 mM)	MN	Dose-related increase of MN for Tradescancia, ND in V. faba
P. sativum	Rodriguez (2011b)	$CrCl_3$ (up to 2000 mg/L)	Comet Assay FCM	DNA damage Clastogenicity G_2/M arrest

Table 1. Literature survey of Cr genotoxic effect in plants.

Rodriguez et al. (2011) showed that flow cytometry (FCM) and Comet assays provided accurate and sensitive biomarkers of the DNA damage endpoint, detecting significant changes in both roots and leaves of plants exposed to Cr(VI) (Figure 2). The level of DNA damage observed in roots was significantly higher than that of the leaves. Roots had direct contact with the metals and it is known that in most cases, this organ acts like a barrier against metal translocation, which might justify why the higher level of DNA damage observed was in roots.

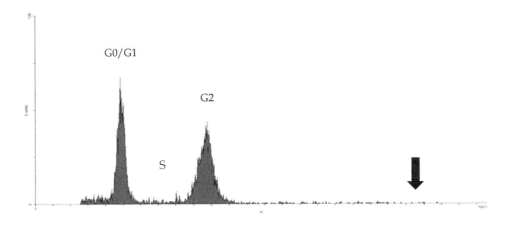

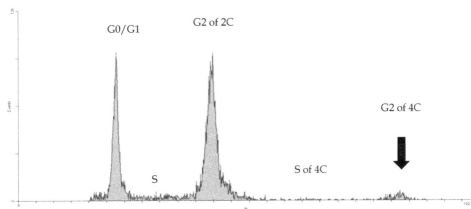

Fig. 2. Fluorescence histograms of control (a) and 2000mg/l (b) Cr (VI) exposed pea roots. Values are given in channels (X axis) and n° of events (Y axis). The arrow indicates the position of extra peak in the bottom histogram (not present in control, above) (adapted from Rodriguez et al 2011b).

Also, under the same Cr(VI) conditions, 40% of the individuals analyzed suffered polyploidization having both 2C and 4C levels. Rodriguez et al (2011) also demonstrated that the clastogenic data provided FCM supported those of Comet assays, and that both tools complemented each other in genotoxicity evaluations.

For the putative cytostatic effects induced by Cr(VI), we have recently showed that Cr(VI) (up to 2000 mg/L) induced few changes in *Pisum sativum* leaves and roots cell cycle progression and that these changes were dependent on the organ and on Cr(VI) concentration: pea leaves showed no significant variations in either cell cycle dynamics. Contrarily, in roots, exposure to 2000 mg/L resulted in cell cycle arrest at the G_2/M

checkpoint (Rodriguez et al., 2011). This may support that an arrest of the cell cycle at this checkpoint occurs when DNA synthesis has been compromised, to give cells extra time to either repair the damage (O'Connell and Cimprich, 2005) or activate an apoptosis-like program (Figure 3). In some cases though, cells might continue with proliferation without completing the damage repair (Carballo et al., 2006). Moreover, the evaluation of MSI helped to explain why, despite that significant DNA damage was detected in lower dosages, only at the maximum dosage an arrest at the cell cycle was observed, since signs of MSI could only be observed at that dosage (Rodriguez 2011).

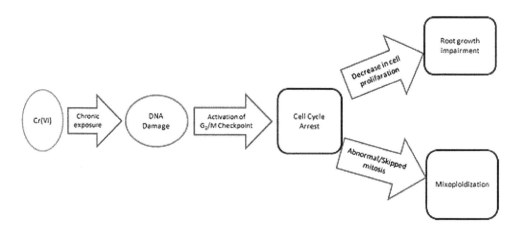

Fig. 3. Cr(VI), after inducing a critical level of DNA damage, leads to malfunction of the DNA repair system, which might in turn induce problems with the cell cycle/division machinery, causing arrest and in extreme cases, polyploidization (from Rodriguez et al., 2011b).

As demonstrated by the authors, flow cytometry (FCM) is a technique that can easily excel in genotoxic studies, allying high analytical speed with multiparametric analysis (with a single analysis can provide information on variations in DNA content and polyploidization, variations in cell cycle dynamics and also, DNA damage). In plants, FCM has been demonstrated to detected differences in DNA content as small as 1% (Pfosser et al., 1995), chromosome aberration in wheat-rye lines exposed to aluminium (Rayburn and Wetzel, 2002), DNA damage in lettuce plants exposed to Cd (Monteiro et al., 2010) and cell cycle arrest in *A .cepa* exposed to X-ray radiation (Carballo et al., 2006).

Another good technique for assessing genotoxicity is the Comet assay, which is a versatile and sensitive method for measuring single- and double-strand breaks in DNA (Collins et al., 2008). The simplicity inherent to sample preparation and the relatively small number of cells/nuclei analysis required to obtained robust results (Hattab et al., 2009a), the later which can be automated further reducing the time need to obtain results, can be accounted as the reasons for the dramatic increase of Comet assays application in genotoxicity studies. In plants, Comet assay has been proved to be very useful to study genotoxicity of heavy metals (e.g Hattab et al., 2009a; Gichner et al., 2006; 2008a).

5. Lead: The element

Lead (Pb) is a silvery-white highly malleable metal, with a low melting point and high density. Pb has had many applications since its discovery: The Egyptians used grounded lead ore as eyeliner with therapeutic proprieties; Pb based pigments were used as part of yellow red and white paint; in ancient Rome, Pb was used to build pipes for water transportation and not so long ago, tetraethyl lead was used in petrol fuels. Nowadays, this metal remains a major constituent of most batteries used in automobiles, is used in projectiles for firearms, and molten Pb is used as a coolant.

Releases of lead in the environment can occur naturally from the mobilization of Pb from the Earth's crust and mantle, such as volcanic activity and the weathering of rocks. However, these releases are very rare and the most significant sources of Pb discharge are those originated by anthropogenic activities.

Some of the most influential sources of Pb pollution are lead impurities in raw materials such as fossil fuels and other extracted and treated metals, mining, releases from incineration and installations for municipal waste, open burning and the mobilization of historical Pb releases previously deposited in soils, sediments and wastes. From these sources this pollutant can be transported thousands of kilometres through the air (burned fuel and air-borne particles like fly ash) and by rivers and oceans (discharges from industries and leakage from residues).

In the atmosphere, lead will deposit on surfaces or exist as a component of atmospheric particles. In the atmosphere, lead exists primarily as lead compounds. The residence time ranges from hours to weeks. In the aquatic environment, lead can occur in ionic form (highly mobile and bio-available), organic complexes with dissolved humus materials (binding is rather strong and limits availability), attached to colloidal particles such as iron oxide (strongly bound and less mobile when available in this form than as free ions) or to solid particles of clay or dead remains of organisms (very limited mobility and availability).

6. Lead: Uptake and assimilation by plants

The speciation of lead differs whether it is in fresh water, seawater or soil. In fresh water lead primarily exists as the divalent cation (Pb^{2+}) under acidic conditions, and forms $PbCO_3$ and $Pb(OH)_2$ under alkaline conditions. Lead speciation in seawater is a function of chloride concentration and the primary species are $PbCl_3^- > PbCO_3 > PbCl_2 > PbCl^+ >$ and $Pb(OH)^+$.

In soil, lead is generally not very mobile. The downward movement of elemental lead and inorganic lead compounds from soil to groundwater by leaching is very slow under most natural conditions. Clays, silts, iron and manganese oxides, and soil organic matter can bind lead and other metals electrostatically (cation exchange) as well as chemically (specific adsorption). Biotic factors like soil pH, content of humic acids and amount of organic matter influence the content and mobility of lead in soils. Despite the fact that lead is not very mobile in soil, lead may enter surface waters as a result of erosion of lead-containing soil particles. All these factors will influence the bioavailability of Pb and thus the toxicity level of this heavy metal.

To become metabolized by plants, elements need to be transported, at some point, through the plasma membrane of the roots. (Kučera et al., 2008) reported that once in contact with plants, Pb was transported by CPx-type ATPases, a subgroup of P-type ATPases, that pump essential and non-essential metals such as Cu^{2+}, Zn^{2+}, Cd^{2+}, and Pb^{2+} across the plasma membrane.

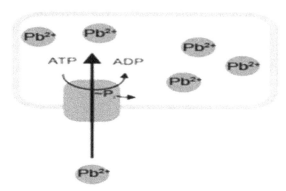

Fig. 4. Pb transport by CPx-type ATPases (adapted from the model of Dr. Mathias Lübben)

In soil, lead is generally not very mobile. The downward movement of elemental lead and inorganic lead compounds from soil to groundwater by leaching is very slow under most natural conditions. Clays, silts, iron and manganese oxides, and soil organic matter can bind lead and other metals electrostatically (cation exchange) as well as chemically (specific adsorption). Biotic factors like soil pH, content of humic acids and amount of organic matter influence the content and mobility of lead in soils. Despite the fact that lead is not very mobile in soil, lead may enter surface waters as a result of erosion of lead-containing soil particles. All these factors will influence the bioavailability of Pb and thus the toxicity level of this heavy metal.

To become metabolized by plants, elements need to be transported, at some point, through the plasma membrane of the roots. (Kučera et al., 2008) reported that once in contact with plants, Pb was transported by CPx-type ATPases, a subgroup of P-type ATPases, that pump essential and non-essential metals such as Cu^{2+}, Zn^{2+}, Cd^{2+}, and Pb^{2+} across the plasma membrane.

Plants absorb Pb usually accumulating it in the roots (Carruyo et al., 2008; Hanc et al., 2009), acting like a natural barrier. Although, a small portion can also be translocated upwards to stems, leaves (Hanc et al., 2009) and seeds being the increase level, directly proportional to the amount of exogenous lead.

Authors have studied the effect of pH variation in Pb uptake, in different plant species: in low pH soils (3.9), increased mobility of lead was observed, resulting in higher uptake (Ernst et al., 2000); Gorlach et al. (1990), working Italian ryegrass found that with increasing soil pH (3.9–6.7), Pb uptake was reduced. Also, and in addition to soil factors, the species and genotype also dictate Pb's uptake and accumulation.

Once inside the root cortex, Pb moves in the apoplastic space, using the transpiration conductive system (Wierzbicka, 1999; Hanc et al., 2009)). It can also bypass the endodermis and gain symplastic access in the young root zone and in sites of lateral root initiation (Eun et al., 2000). Pb has been shown to enter and move within the cytoplasm and proteins mediating cross-membrane movement of Pb have been identified (Kerper and Hinkle, 1997; Arazi et al., 1999). Most of the Pb absorbed by roots exists as extracellular precipitate (as phosphate and carbonate)or is bound to ion exchangeable sites in the cell walls (Sahi et al., 2002). The unbound Pb is moved through Ca channels accumulating near the endodermis (Huang and Cunningham, 1996; Antosiewicz, 2005). Depending on the plant species exposed, different cellular types can be used to store Pb: in wheat, Pb is fixed to the cell wall of roots but it can be removed as a complex using citric acid (Varga et al., 1997). Peralta-Videa et al. (2009) on the other hand discuss the accumulation of Pb in the phloem tissues of *Prosopis* sp. associated with *Glomus deserticola,* suggesting that it was transported to the leaves and returned through the phloem to the plant organs.

7. Lead: Phytotoxicity

7.1 General effect

The first reported uses of lead date back to 4000 BC, and toxicological effects have been linked to lead since antiquity. Lead is known to bioaccumulate in most organisms, whereas it is generally not biomagnified up the food web.

Pb its known to negatively affect some of the most classical endpoints of plant toxicity like germination rate, growth and dry mass of roots and shoots (Ekmekçi et al., 2009; Munzuroglu and Geckil, 2002;). In general, effects are more pronounced at higher concentrations and durations. In some cases, lower concentrations stimulate metabolic processes and enzymes involved. The major processes affected are seed germination, seedling growth (shoot and root growth), photosynthesis, plant water status, mineral nutrition, and enzymatic activities. Visible symptoms include chlorotic spots, necrotic lesions in leaf surface, senescence of leaf and stunted growth. Germination of seeds is drastically affected at higher concentrations. Development and growth of root and shoot in seedling stage are also affected, roots being more sensitive.

Lead reduced the uptake and transport of nutrients in plants, such as Ca, Fe, Mg, Mn, P and Zn, by blocking the entry or binding of the ions to ion-carriers making them unavailable for uptake and transport from roots to leaves (Xiong, 1997). This in turn affects several physiological and biochemical processes, among which photosynthesis is one of the most affected.

7.2 Photosynthesis

Photosynthesis is one of the processes most sensitive to lead: the substitution of the central atom of chlorophyll, magnesium, by lead in vivo prevents photosynthetic light-harvesting in the affected chlorophyll molecules, resulting in a breakdown of photosynthesis (Küpper et al., 1996). Higher concentrations of lead significantly affected plant water status causing water deficit.

The deleterious effects of this metal in several physiological parameters have been addressed in several species: John et al. (2009) found in *Brassica juncea* exposed to this metal, growth impairment and decrease in pigments content; Kosobrukhov et al. (2004) working with *Platango major* showed that Pb can affect g_s, pigment content, and light and dark reactions; Bibi and Hussain (2005) demonstrated that the A, E and g_s of *Vigna mungo* plants were significantly affected when exposed to Pb. The total chlorophyll content and relative content proportion of Chl a and b were reduced, through inhibition of chlorophyll biosynthesis (Ernst et al., 2000; Van Assche and Clijsters, 1990; Sengar and Pandey, 1996). Cenkci et al. (2010) found that the content in carotenoids was less affected than chlorophylls by Pb and suggested that this was so because carotenoids protect chlorophyll from photo-oxidative destruction and therefore, a reduction in carotenoids could have a serious consequence on chlorophyll pigments.Limitation of photosynthesis by reduced activity of Calvin cycle enzymes, e.g. RuBisCO activity was reported for several plant species exposed to Pb (Vojtěchová and Leblová, 1991; Moustakas et al., 1994). Lee and Roh (2003) found that exposure to Cd induced significant decrease in RuBisCO activity which was associated to the amount of RuBisCO protein; this might be a hint to the decrease in RuBisCO activity observed with Pb exposure, it is possible that Pb as Cd cause a decrease in RuBisCO protein. More recently, Bah et al. (2010) proved that Pb caused the up-regulation of carbohydrate metabolic pathway enzymes; APX and GRSF; RuBisCO activase, Mg-protoporphyrin IX chelatase, fructokinase, a chloroplast precursor and plastocyanin suggests. With those results, the authors concluded that what was observed was part of a strategy to cope with Pb toxicity, by increasing carbohydrate metabolism (fruktokinase), photosynthesis (RuBisCO activase, Mg-protoporphyrin chelatase and plastocyanin) and defense response (APX and GRSF). They also concluded that despite that the strategy was responsible for the high tolerance of *T. angustifolia* to Pb toxicity, this had a high energetic cost.

Transpiration intensity, osmotic pressure of cell sap, water potential of xylem, and relative water content were significantly reduced after 24 and 48h of exposure to Pb (Parys et al., 1998). Lead also reduces the size of stomata but increases their number and diffusion resistance.

The mechanism(s) of this metal toxicity on photosynthesis is still a matter of speculations, this may be partly due to the differences in experimental design, but it almost certainly involves electron transport in light reactions and enzyme activity in the dark reactions (Romanowska et al., 2006). Despite of the fact that the mechanism by which Pb affects the photosynthetic apparatus is unclear, evidence indicates that this metal causes severe effect to the photosynthetic status of plants and thus, it of vital importance to carry studies to better understand Pb's toxicity.

Similarly to the studies with Cr(VI), Rodriguez (2011) evaluated the Pb -induced toxicity on the photosynthetic status of *Pisum sativum* plants. The endpoints measured involved gas exchange, Calvin cycle enzymes activity, amount of soluble sugars and starch, pigment content and fluorescence emission. Moreover, chloroplast structure and functional status variation as function of Pb toxicity were also demonstrated in pea leaves by FCM.

7.3 Genotoxicity

The chemical form of Pb only affects lead transport from the medium into the plants and all forms had similar effects on mitosis. The iodides had a greater mutagenic effect than the nitrates, perhaps because the latter dissolved completely in the solution and were supplied as ions, rather than molecules as in the cases of the iodides (Radecki et al., 1989).

Pb toxicity has been linked to carcinogenicity and the genotoxic effects of this metal have been studied thoroughly in animals and humans. Nevertheless, data related to the mutagenic, clastogenic and carcinogenic properties of inorganic lead compounds are still conflicting (García-Lestón et al., 2010). Hartwig et al. (1990) working with V79 Chinese hamster cells exposed to Pb and UV radiation concluded that Pb alone did not induced DNA damage but magnified that caused by UV rays, this they said, was due to Pb interference with the repair machinery. This might be due to Pb ability to substitute calcium and/or zinc in enzymes involved in DNA processing and repair leading to an inhibition of DNA repair and an enhancement in the genotoxicity when combined with other DNA damaging agents (García-Lestón et al., 2010). The major mechanisms putatively involved in Pb genotoxicity are summarized in Figure 5.

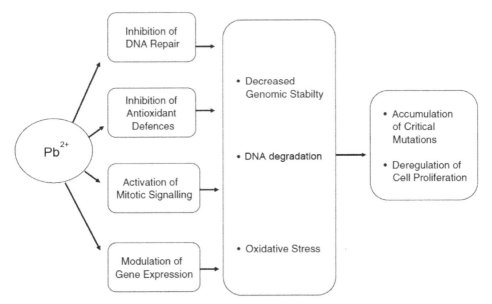

Fig. 5. Major mechanism of Pb genotoxicity. Adapted from Beyersmann and Hartwig (2008).

Valverde et al. (2001) demonstrated that despite that Pb did not cause DNA damage in in vitro DNA, production of lipid peroxidation and an increase in free radical levels were observed, suggesting that Pb exposure cause genotoxicity and carcinogenicity by indirect interactions, such as oxidative stress. These investigations support the current thesis stating that the way of action of Pb might be through ROS formation and interference with the DNA repair mechanism (Beyersmann and Hartwig, 2008), rather than a direct interaction with DNA as it is seen with Cr.

Animal cell proliferation has also been demonstrated to be sometimes affected by Pb exposure, by increasing proliferative lesions in the kidney, below cytotoxic concentrations. This stimulation indicates that genotoxicity and accelerated growth stimuli may act in concert in lead-induced carcinogenicity in mammals (Beyersmann and Hartwig, 2008).

In plants and despite of the importance of Pb pollution and risk associated to the environment, the mechanism and effects of Pb toxicity are far less known than in animals. Of what is known, most of the Pb absorbed remains in the root with only a small fraction being translocated to the shoots (Patra et al., 2004). There, Pb has been demonstrated to cause chromosome aberration (Carruyo et al., 2008) in *A. cepa*; DNA degradation in lupin and tobacco (Gichner et al., 2008b, Rucinska et al., 2004) and genomic instability in turnips (Cenkci et al., 2010). Lead nitrate proved to be a weak mutagen but owing to its high toxicity had a synergistic effect in combination with ionizing radiation in some populations (Patra et al., 2004). Lead in particular, has been demonstrated to increase Comets formation, thus having genotoxic effects, at short term exposure, in tobacco (Gichner et al., 2008b) and lupin (Rucinska et al., 2004). Pb genotoxicity studies in plants are summarized in Table 2.

Recently our group demonstrated that leaves from Pb-exposed plants showed a slight increase in DNA degradation at the highest tested concentration, while in roots, significant changes in cell cycle dynamics were observed at G_0/G_1 and G_2. In these roots, significant damages of DNA were shown by increases of tail moment (TM) and of full peak coefficient of variation (FPCV). The authors suggested that Pb induced a blockage of cell cycle at the G_2/M checkpoint due to severe degradation of the DNA (Rodriguez 2011).

Species	Reference	Dose	Technique	Effects
P. sativum	(Gabara et al., 1995)	10^{-4} M	DNA synthesis	Diminished DNA synthesis
V. faba	(Chang-qun and Huan-xiao, 1995)	Pb^{2+} (NR)	cytogenetic	Mitotic stage shortened and interphase prolonged
A. cepa	(Rank and Nielsen, 1998)	Wastewater Sludges	cytogenetic	Anaphase-Telophase chromosome aberration
H. vulgare *C. sativum* *A. cepa*	(Bhowmik, 2000)	$Pb(NO_3)_2$ (0.001 to 1 mg/Kg)	cytogenetic	Redution of mitotic index, increase of chromosomal aberration, Polyploidy
A. thaliana (transgenic)	(Kovalchuk and Yao, 2011)	Pb^{2+} (0.002 to 0.83 mg/L)	Trasngenic plant reporter gene	Increase in the mutation frequency
L. luteus	(Rucinska et al., 2004)	$Pb(NO_3)_2$ (150 and 350 mg/l)	Comet assay	DNA damage
N. tabacum	(Gichner et al., 2008b)	Pb^{2+} (200 µM to 0.4 mM)	Comet Assay	DNA damage
B. rapa	(Cenkci et al., 2010)		RAPD	Genomic template instability
L. sativa	(Ritambhara and Girjesh, 2010)	$Pb(NO_3)_2$ (25 to 300 ppm)	cytogenetic	Abnormal chromosome migration
V.faba	(Shahid et al., 2011))	$Pb(NO_3)_2$ (5 µM)	Cytogenetic	MN and mitotic index
P sativum	Rodriguez 2011	$Pb(NO_3)_2$ (up to 2000 mg/L)	Comet Assay FCM	DNA damage Clastogenicity G_2/M arrest

Table 2. Literature survey of Pb genotoxic effect in plants.

8. Acknowledgement

This work was supported by Portuguese Foundation for Science and Technology (FCT) FCT/PTDC/AAC-AMB/112804/2009, BioRem: Integration of multiple Biomarkers of toxicity in an assay of phytoremediation in contaminated sites. FCT is also thanked for the fellowships of E. Rodriguez (SFRH/BD/27467/2006).

9. References

Antosiewicz DM. 2005. Study of calcium-dependent lead-tolerance on plants differing in their level of Ca-deficiency tolerance. *Environmental Pollution*, 134: 23-34.

Appenroth KJ, Bischoff M, Gabrys H, Stoeckel J, Swartz HM, Walczak T & Winnefeld K. 2000. Kinetics of chromium(V) formation and reduction in fronds of the duckweed *Spirodela polyrhiza* -- a low frequency EPR study. *Journal of Inorganic Biochemistry*, 78: 235-242.

Arazi T, Sunkar R, Kaplan B & Fromm H. 1999. A tobacco plasma membrane calmodulin-binding transporter confers Ni2+ tolerance and Pb2+ hypersensitivity in transgenic plants. *The Plant Journal*, 20: 171-182.

Babula P, Adam V, Opatrilova R, Zehnalek J, Havel L & Kizek R. 2008. Uncommon heavy metals, metalloids and their plant toxicity: a review. *Environmental Chemistry Letters*, 6: 189-213.

Bah AM, Sun HY, Chen F, Zhou J, Dai HX, Zhang GP & Wu FB. 2010. Comparative proteomic analysis of Typha angustifolia leaf under chromium, cadmium and lead stress. *Journal of Hazardous Materials*, 184: 191-203.

Beyersmann D & Hartwig A. 2008. Carcinogenic metal compounds: recent insight into molecular and cellular mechanisms. *Archives of Toxicology*, 82: 493-512.

Bhowmik MK. 2000. *Cytotoxic effects of lead compounds on plant systems.* , PhD, University of Calcutta, Calcutta, India.

Bibi M & Hussain M. 2005. Effect of copper and lead on photosynthesis and plant pigments in black gram [*Vigna mungo* (L.) Hepper]. *Bulletin of Environmental Contamination and Toxicology*, 74: 1126-1133.

Carballo JA, Pincheira J & De la Torre C. 2006. The G2 checkpoint activated by DNA damage does not prevent genome instability in plant cells. *Biological Research*, 39: 331-340.

Carruyo I, Fernández Y, Marcano L, Montiel X & Torrealba Z. 2008. Correlation of Toxicity with Lead Content in Root Tip Cells (*Allium cepa* L.). *Biological Trace Element Research*, 125: 276-285.

Cenkci S, Cigerci IH, Yildiz M, Ozay C, Bozdag A & Terzi H. 2010. Lead contamination reduces chlorophyll biosynthesis and genomic template stability in Brassica rapa L. *Environmental and Experimental Botany*, 67: 467-473.

Chandra S, Chauhan LKS, Pande PN & Gupta SK. 2004. Cytogenetic effects of leachates from tannery solid waste on the somatic cells of Vicia faba. *Environmental Toxicology*, 19: 129-133.

Chang-qun D & Huan-xiao W. 1995. Cytogenetical toxical effects of heavy metals on *Vicia faba* and inquires into the *Vicia*-micronucleus. *Acta Botanica Sinica*, 37: 10.

Chen N, Kanazawa S, Horiguchi T & Chen N. 2001. Effect of chromium on some enzyme activities in the wheat rhizosphere. *Soil Microorganims*, 55: 7.

Citterio S, Aina R, Labra M, Ghiani A, Fumagalli P, Sgorbati S & Santagostino A. 2002. Soil genotoxicity assessment: A new strategy based on biomolecular tools and plant bioindicators. *Environmental Science & Technology*, 36: 2748-2753.

Citterio S, Santagostino A, Fumagalli P, Prato N, Ranalli P & Sgorbati S. 2003. Heavy metal tolerance and accumulation of Cd, Cr and Ni by Cannabis sativa L. *Plant and Soil*, 256: 243-252.

Collins AR, Oscoz AA, Brunborg G, Gaivao I, Giovannelli L, Kruszewski M, Smith CC & Stetina R. 2008. The comet assay: topical issues. *Mutagenesis*, 23: 143-151.

Dhir B, Sharmila P, Saradhi PP & Nasim SA. 2009. Physiological and antioxidant responses of Salvinia natans exposed to chromium-rich wastewater. *Ecotoxicology and Environmental Safety*, 72: 1790-1797.

Duffus, J.H., 2002. "Heavy metals" - A meaningless term? (IUPAC Technical Report). Pure Appl Chem 74, 793-807.

Ekmekçi Y, Tanyolaç D & Ayhan B. 2009. A crop tolerating oxidative stress induced by excess lead: maize. *Acta Physiologiae Plantarum*, 31: 319-330.

Ernst WHO, Nelissen HJM & Ten Bookum WM. 2000. Combination toxicology of metal-enriched soils: physiological responses of a Zn- and Cd-resistant ecotype of *Silene vulgaris* on polymetallic soils. *Environmental and Experimental Botany*, 43: 55-71.

Eun SO, Youn HS & Lee Y. 2000. Lead disturbs microtubule organization in the root meristem of Zea mays. *Physiologia Plantarum*, 110: 357-365.

Gabara B, Krajewska M & Stecka E. 1995. Calcium Effect on Number, Dimension and Activity of Nucleoli in Cortex Cells of Pea (Pisum-Sativum L) Roots after Treatment with Heavy-Metals. *Plant Science*, 111: 153-161.

García-Lestón J, Méndez J, Pásaro E & Laffon B. 2010. Genotoxic effects of lead: An updated review. *Environment International*, 36: 623-636.

Gardea-Torresdey JL, de la Rosa G, Peralta-Videa JR, Montes M, Cruz-Jimenez G & Cano-Aguilera I. 2005. Differential Uptake and Transport of Trivalent and Hexavalent Chromium by Tumbleweed (<i>Salsola kali</i>). *Archives of Environmental Contamination and Toxicology*, 48: 225-232.

Gichner T, Patkova Z, Szakova J & Demnerova K. 2006. Toxicity and DNA damage in tobacco and potato plants growing on soil polluted with heavy metals. *Ecotoxicology and Environmental Safety*, 65: 420-426.

Gichner T, Patkova Z, Szakova J, Znidar I & Mukherjee A. 2008a. DNA damage in potato plants induced by cadmium, ethyl methanesulphonate and gamma-rays. *Environmental and Experimental Botany*, 62: 113-119.

Gichner T, Znidar I & Szakova J. 2008b. Evaluation of DNA damage and mutagenicity induced by lead in tobacco plants. *Mutation Research-Genetic Toxicology and Environmental Mutagenesis*, 652: 186-190.

Gyuricza V, Fodor F & Szigeti Z. 2010. Phytotoxic Effects of Heavy Metal Contaminated Soil Reveal Limitations of Extract-Based Ecotoxicological Tests. *Water Air and Soil Pollution*, 210: 113-122.

Hanc A, Baralkiewicz D, Piechalak A, Tomaszewska B, Wagner B & Bulska E. 2009. An analysis of long-distance root to leaf transport of lead in Pisum sativum plants by laser ablation-ICP-MS. *International Journal of Environmental Analytical Chemistry*, 89: 651-659.

Hattab S, Chouba L, Ben Kheder M, Mahouachi T & Boussetta H. 2009a. Cadmium- and copper-induced DNA damage in Pisum sativum roots and leaves as determined by the Comet assay. *Plant Biosystems,* 143: S6-S11.

Hattab S, Dridi B, Chouba L, Kheder MB & Bousetta H. 2009b. Photosynthesis and growth responses of pea Pisum sativum L. under heavy metals stress. *Journal of Environmental Sciences-China,* 21: 1552-1556.

Henriques FS. 2010. Changes in biomass and photosynthetic parameters of tomato plants exposed to trivalent and hexavalent chromium. *Biologia Plantarum,* 54: 583-586.

Hewitt EJ. 1953. Metal Interrelationships in Plant Nutrition. *Journal of Experimental Botany,* 4: 59-64.

Hong S, Candelone J-P, Patterson CC & Boutron CF. 1994. Greenland Ice Evidence of Hemispheric Lead Pollution Two Millennia Ago by Greek and Roman Civilizations. *Science,* 265: 1841-1843.

Huang JW & Cunningham SD. 1996. Lead phytoextraction: species variation in lead uptake and translocation. *New Phytologist,* 134: 75-84.

Iqbal M, Saeeda S & Shafiq M. 2001. Effects of chromium on an important arid tree (*Caesalpinia pulcherrima*) of Karachi city, Pakistan. *Ekol Bratislava,* 20: 8.

John R, Ahmad P, Gadgil K & Sharma S. 2009. Heavy metal toxicity: Effect on plant growth, biochemical parameters and metal accumulation by Brassica juncea L. *International Journal of Plant Production,* 3: -.

Juarez AB, Barsanti L, Passarelli V, Evangelista V, Vesentini N, Conforti V & Gualtieri P. 2008. In vivo microspectroscopy monitoring of chromium effects on the photosynthetic and photoreceptive apparatus of Eudorina unicocca and Chlorella kessleri. *Journal of Environmental Monitoring,* 10: 1313-1318.

Kerper LE & Hinkle PM. 1997. Cellular Uptake of Lead Is Activated by Depletion of Intracellular Calcium Stores. *Journal of Biological Chemistry,* 272: 8346-8352.

Knasmuller S, Gottmann E, Steinkellner H, Fomin A, Pickl C, Paschke A, God R & Kundi M. 1998. Detection of genotoxic effects of heavy metal contaminated soils with plant bioassays. *Mutation Research-Genetic Toxicology and Environmental Mutagenesis,* 420: 37-48.

Koppen G & Verschaeve L. 1996. The alkaline comet test on plant cells: A new genotoxicity test for DNA strand breaks in Vicia faba root cells. *Mutation Research-Environmental Mutagenesis and Related Subjects,* 360: 193-200.

Kosobrukhov A, Knyazeva I & Mudrik V. 2004. *Plantago major plants responses to increase content of lead in soil: Growth and photosynthesis,* Dordrecht, PAYS-BAS, Springer.

Kovalchuk I & Yao YL. 2011. Abiotic stress leads to somatic and heritable changes in homologous recombination frequency, point mutation frequency and microsatellite stability in Arabidopsis plants. *Mutation Research-Fundamental and Molecular Mechanisms of Mutagenesis,* 707: 61-66.

Kučera T, Horáková H & Šonská A. 2008. Toxic metal ions in photoautotrophic organisms. *Photosynthetica,* 46: 481-489.

Küpper H, Küpper F & Spiller M. 1996. Environmental relevance of heavy metal-substituted chlorophylls using the example of water plants. *Journal of Experimental Botany,* 47: 259-266.

Labra M, Di Fabio T, Grassi F, Regondi SMG, Bracale M, Vannini C & Agradi E. 2003. AFLP analysis as biomarker of exposure to organic and inorganic genotoxic substances in plants. *Chemosphere,* 52: 1183-1188.

Labra M, Grassi F, Imazio S, Di Fabio T, Citterio S, Sgorbati S & Agradi E. 2004. Genetic and DNA-methylation changes induced by potassium dichromate in *Brassica napus* L. *Chemosphere,* 54: 1049-1058.

Lee K & Roh K. 2003. Influence of cadmium on rubisco activation in *Canavalia ensiformis* L. leaves. *Biotechnology and Bioprocess Engineering,* 8: 94-100.

Lopez-Luna J, Gonzalez-Chavez MC, Esparza-Garcia FJ & Rodriguez-Vazquez R. 2009. Toxicity assessment of soil amended with tannery sludge, trivalent chromium and hexavalent chromium, using wheat, oat and sorghum plants. *Journal of Hazardous Materials,* 163: 829-834.

Matsumoto ST, Mantovani MS, Malaguttii MIA, Dias AU, Fonseca IC & Marin-Morales MA. 2006. Genotoxicity and mutagenicity of water contaminated with tannery effluents, as evaluated by the micronucleus test and comet assay using the fish *Oreochromis niloticus* and chromosome aberrations in onion root-tips. *Genetics and Molecular Biology,* 29: 148-158.

McConnell JR & Edwards R. 2008. Coal burning leaves toxic heavy metal legacy in the Arctic. *Proceedings of the National Academy of Sciences,* 105: 12140–12144.

Mittler R. 2002. Oxidative stress, antioxidants and stress tolerance. *Trends in Plant Science,* 7: 405-410.

Monteiro MS, Rodriguez E, Loureiro J, Mann RM, Soares AMVM & Santos C. 2010. Flow cytometric assessment of Cd genotoxicity in three plants with different metal accumulation and detoxification capacities. *Ecotoxicology and Environmental Safety,* 73: 1231-1237.

Munzuroglu O & Geckil H. 2002. Effects of Metals on Seed Germination, Root Elongation, and Coleoptile and Hypocotyl Growth in *Triticum aestivum* and *Cucumis sativus*. *Archives of Environmental Contamination and Toxicology,* 43: 203-213.

Murozumi M, Chow TJ & Patterson C. 1969. Chemical concentrations of pollutant lead aerosols, terrestrial dusts and sea salts in Greenland and Antarctic snow strata. *Geochimica Et Cosmochimica Acta,* 33: 1247-1294.

Nriagu JO. 1996. A History of Global Metal Pollution. *Science,* 272: 223.

O'Brien TJ, Fornsaglio JL, Ceryak S & Patierno SR. 2002. Effects of hexavalent chromium on the survival and cell cycle distribution of DNA repair-deficient *S. cerevisiae*. *DNA Repair,* 1: 617-627

Oliveira, H., Loureiro, J., Filipe, L., Santos, C., Ramalho-Santos, J., Sousa & M. Pereira, M.d.L., 2006. Flow cytometry evaluation of lead and cadmium effects on mouse spermatogenesis. Reprod Toxicol 22, 529.

Otto, F.J. & Oldiges H. 1980. Flow cytogenetic studies in chromosomes and whole cells for the detection of clastogenic effects. Cytometry 1, 13-17.

Pandey V, Dixit V & Shyam R. 2009. Chromium effect on ROS generation and detoxification in pea (*Pisum sativum*) leaf chloroplasts. *Protoplasma,* 236: 85-95.

Patra M, Bhowmik N, Bandopadhyay B & Sharma A. 2004. Comparison of mercury, lead and arsenic with respect to genotoxic effects on plant systems and the development of genetic tolerance. *Environmental and Experimental Botany,* 52: 199-223.

Peralta-Videa JR, Lopez ML, Narayan M, Saupe G & Gardea-Torresdey J. 2009. The biochemistry of environmental heavy metal uptake by plants: Implications for the food chain. *International Journal of Biochemistry & Cell Biology*, 41: 1665-1677.

Pfosser M, Amon A, Lelley T & Heberlebors E. 1995. Evaluation of Sensitivity of Flow-Cytometry in Detecting Aneuploidy in Wheat Using Disomic and Ditelosomic Wheat-Rye Addition Lines. *Cytometry*, 21: 387-393.

Prado FE, Prado C, Rodriguez-Montelongo L, Gonzalez JA, Pagano EA, Hilal M. 2010. Uptake of chromium by *Salvinia minima*: Effect on plant growth, leaf respiration and carbohydrate metabolism. *Journal of Hazardous Materials*, 177: 546-553.

Prasad MN, Greger M, Landberg T. 2001. Acacia nilotica L. Bark Removes Toxic Elements from Solution: Corroboration from Toxicity Bioassay Using Salix viminalis L. in Hydroponic System. *International Journal of Phytoremediation*, 3: 289-300.

Rank J & Nielsen MH. 1998. Genotoxicity testing of wastewater sludge using the *Allium cepa* anaphase-telophase chromosome aberration assay. *Mutation Research-Genetic Toxicology and Environmental Mutagenesis*, 418: 113-119.

Rayburn AL & Wetzel JB. 2002. Flow cytometric analyses of intraplant nuclear DNA content variation induced by sticky chromosomes. *Cytometry*, 49: 36-41.

Ritambhara T & Girjesh K. 2010. Genetic loss through heavy metal induced chromosomal stickiness in Grass pea. *Caryologia*, 63: 223-228.

Rodriguez E, Santos C, Lucas E & Pereira MJ. 2011 a. Evaluation of Chromium (Vi) Toxicity to *Chlorella vulgaris* Beijerinck Cultures. *Fresenius Environmental Bulletin*, 20: 334-339.

Rodriguez E, Azevedo R, Fernandes P & Santos C. 2011. b Cr(VI) Induces DNA Damage, Cell Cycle Arrest and Polyploidization: A Flow Cytometric and Comet Assay Study in *Pisum sativum Chemical Research Toxicololgy* 24 (7), 1040–1047

Rodriguez E, Azevedo R, Costa A, Serôdio J & Santos C. 2011c. Chloroplast functionality assessment by flow cytometry. *Photosynthetica* (in press).

Rodriguez E. 2011 Cytotoxicity and genotoxicity of Cr(VI) and Pb in *Pisum sativum*. PhD Thesis, University Aveiro, Portugal.

Rolland F, Baena-Gonzalez E & Sheen J. 2006. Sugar sensing and signalling in plants: Conserved and Novel Mechanisms. *Annual Review of Plant Biology*, 57: 675-709.

Romanowska E, Wróblewska B, Dro, ak A & Siedlecka M. 2006. High light intensity protects photosynthetic apparatus of pea plants against exposure to lead. *Plant Physiology and Biochemistry*, 44: 387-394.

Rucinska R, Sobkowiak R & Gwozdz EA. 2004. Genotoxicity of lead in lupin root cells as evaluated by the comet assay. *Cellular & Molecular Biology Letters*, 9: 519-528.

Sahi SV, Bryant NL, Sharma NC & Singh SR. 2002. Characterization of a Lead Hyperaccumulator Shrub, Sesbania drummondii. *Environmental Science & Technology*, 36: 4676-4680.

Salnikow K & Zhitkovich A. 2007. Genetic and Epigenetic Mechanisms in Metal Carcinogenesis and Cocarcinogenesis: Nickel, Arsenic, and Chromium. *Chemical Research in Toxicology*, 21: 28-44.

Samantaray S, Rout GR & Das P. 1998. Differential nickel tolerance of mung bean (*Vigna radiata L.*) genotypes in nutrient culture. *Agronomie*, 18: 537-544.

Sengar R & Pandey M. 1996. Inhibition of chlorophyll biosynthesis by lead in greening *Pisum sativum* leaf segments. *Biologia Plantarum*, 38: 459-462.

Seregin IV & Ivanov VB. 2001. Physiological aspects of cadmium and lead toxic effects on higher plants. *Russian Journal of Plant Physiology,* 48: 523-544.

Shahid M, Pinelli E, Pourrut B, Silvestre J & Dumat C. 2011. Lead-induced genotoxicity to Vicia faba L. roots in relation with metal cell uptake and initial speciation. *Ecotoxicology and Environmental Safety,* 74: 78-84.

Shanker AK, Cervantes C, Loza-Tavera H & Avudainayagam S. 2005. Chromium toxicity in plants. *Environment International,* 31: 739-753.

Singh A, Misra P & Tandom PK. 2006. Phytotoxicity of chromium in paddy (*Oryza sativa* L.) plants. *Journal of Environmental Biology,* 27: 2.

Subrahmanyam D. 2008. Effects of chromium toxicity on leaf photosynthetic characteristics and oxidative changes in wheat (*Triticum aestivum* L.). *Photosynthetica,* 46: 339-345.

Tiwari KK, Dwivedi S, Singh NK, Rai UN & Tripathi RD. 2009. Chromium (VI) induced phytotoxicity and oxidative stress in pea (*Pisum sativum* L.): Biochemical changes and translocation of essential nutrients. *Journal of Environmental Biology,* 30: 389-394.

Vajpayee P, Sharma SC, Tripathi RD, Rai UN & Yunus M. 1999. Bioaccumulation of chromium and toxicity to photosynthetic pigments, nitrate reductase activity and protein content of *Nelumbo nucifera* Gaertn. *Chemosphere,* 39: 2159-2169.

Van Assche F & Clijsters H. 1990. Effects of metals on enzyme activity in plants. *Plant, Cell & Environment,* 13: 195-206.

Varga A, Záray G, Fodor F & Cseh E. 1997. Study of interaction of iron and lead during their uptake process in wheat roots by total-reflection X-ray fluorescence spectrometry. *Spectrochimica Acta Part B: Atomic Spectroscopy,* 52: 1027-1032.

Vernay P, Gauthier-Moussard C & Hitmi A. 2007. Interaction of bioaccumulation of heavy metal chromium with water relation, mineral nutrition and photosynthesis in developed leaves of Lolium perenne L. *Chemosphere,* 68: 1563-1575.

Vernay P, Gauthier-Moussard C, Jean L, Bordas F, Faure O, Ledoigt G & Hitmi A. 2008. Effect of chromium species on phytochemical and physiological parameters in Datura innoxia. *Chemosphere,* 72: 763-771.

Vojtěchová M & Leblová S. 1991. Uptake of Lead and Cadmium by Maize Seedlings and the Effect of Heavy Metals on the Activity of Phosphoenolpyruvate Carboxylase Isolated from Maize. *Biologia Plantarum,* 33: 386-394.

Waisberg M, Joseph P, Hale B & Beyersmann D. 2003. Molecular and cellular mechanisms of cadmium carcinogenesis. *Toxicology,* 192: 95-117.

Wang HQ. 1999. Clastogenicity of chromium contaminated soil samples evaluated by Vicia root-micronucleus assay. *Mutation Research-Fundamental and Molecular Mechanisms of Mutagenesis,* 426: 147-149.

Wierzbicka M. 1999. Comparison of lead tolerance in *Allium cepa* with other plant species. *Environmental Pollution,* 104: 41-52.

Xiong Z-T. 1997. Bioaccumulation and physiological effects of excess lead in a roadside pioneer species Sonchus oleraceus L. *Environmental Pollution,* 97: 275-279.

Zhang Z, Leonard SS, Wang SW, Vallyathan V, Castranova V & Shi XL. 2001. CR (VI) induces cell growth arrest through hydrogen peroxide-mediated reactions. *Molecular and Cellular Biochemistry,* 222: 77-83.

Part 2

Economic Botany

4

A Mini-Review on Smut Disease of Sugarcane Caused by *Sporisorium scitamineum*

A. Ramesh Sundar, E. Leonard Barnabas, P. Malathi and R. Viswanathan

Plant Pathology section, Sugarcane Breeding Institute (ICAR), Coimbatore, TamilNadu, India

1. Introduction

Smut of sugarcane is caused by the fungus *Ustilago scitmainea* (Sydow, 1924). The first report of the disease incidence came from Natal, South Africa in 1877 as reported by Luthra *et al.*, (1940) and it was speculated to be confined in the eastern hemisphere, until it was reported in Argentina. The disease spread is worldwide covering most of the sugarcane producing areas *viz.* Mauritius, Rhodesia, Indonesia, the islands of Java, Sulawesi and Sumbawa, etc. Until the 1950s, smut was of concern only in Asia, with an outlying population in Argentina. Since then, it has spread through South, Central, East and West Africa, where many of the areas not having focused breeding programme for smut resistance. In the 1970s and 1980s, it expanded to Hawaii, the Caribbean, the mainland USA, Central America and Southern Brazil. These outbreaks prompted a great deal of experimental work on sugarcane smut (Heinz, 1987). Subsequently, the occurrence of sugarcane smut in Morocco (Akalach, 1994) and Iran (Banihashemi, 1995) was established. The occurrence, prevalence and importance of the sugarcane smut pathogen have been highlighted by Antoine as early as 1961. The incidence of the disease was widespread covering several countries in East Africa, the Pacific and the Caribbean islands, wherein a severe outbreak of the disease resulted in devastating loss to the sugarcane plantations. Lovick (1978) comprehensively reviewed on various aspects of sugarcane smut *viz.* symptoms, yield reduction, causal organism, physiological races of the smut fungus, epidemiology, host resistance and management. It was reported that severe smut outbreak in the Caribbean has created an impact amongst cane growers and sugar industry.

The most recognizable diagnostic feature of sugarcane infected with smut is the emergence of a long, elongated whip. The whip morphology differs from short to long, twisted, multiple whips etc. (Fig.1). Affected sugarcane plants may tiller profusely with spindly and more erect shoots with small narrow leaves (i.e., the cane appears ''grass-like'') with poor cane formation (Fig.2). Others symptoms are leaf and stem galls, and bud proliferation (Fig.3). The disease can cause significant losses in cane tonnage and juice quality; its development and severity depend on the environmental conditions and the resistance of the sugarcane varieties. Successful management of smut in sugarcane relies more on exploiting host resistance. To enhance smut resistance in commercial hybrids, intensive breeding programs should be formulated by involving exotic clones as source of resistance from germplasm exchange. In this connection, identifying resistance genes in the wild *Saccharum spontaneum* and bringing up of newer resistant clones along with an efficient screening program could significantly reduce the disease incidence. Singh *et al.*, (2005) concluded that ITS-based probing could not bring out much of variability among *S. scitamineum* isolates in South Africa and suggested that IGS-based studies could possibly discriminate genetic variability amongst the isolates representing different sugarcane growing regions of the world. Raboin *et al.*, (2007) hypothesized that in sugarcane smut, pathogenic variability is greater in Asian countries, where a high level of genetic variation in *S. scitamineum* is reported. With the advent of more precise molecular tools, it is now possible to understand better - pathogen variability *vis a vis* host resistance, that would augment well for successful disease management in the future.

Fig. 1. **Different forms of whip morphology in smut infected sugarcane.** a) Long whip. b) Closed whip. c) Twisted whip. d) Short whip. e) Multiple whips.

Fig. 2. **Smut infected clump.** Characteristic symptoms of profuse tillering and poor cane formation (left) as compared to healthy canes (right).

Fig. 3. **Unusual symptoms due to smut infection.** a) Apical deformity. b) Floral infection. c) Malformed spindle. d) Bud proliferation.

2. Distribution

Smut disease of sugarcane can cause considerable yield losses and reduction in cane quality (Ferreira & Comstock, 1989). The disease is sometimes referred to as "culmicolous" smut of sugarcane, because it affects the stalk of the cane. Smut disease resulted in significant yield losses in sugarcane production and was reported to be distributed all over the sugarcane growing areas in China (Huang *et al.*, 2004). At one time or another, sugarcane smut has been important in nearly every sugarcane growing country in the world. Australia is a major exception, since the disease was initially present only in Western Australia, a minor production area. The disease was first reported in Australia in the Ord River Irrigation Area (ORIA) in 1998. The most likely source of this infection was thought to be windblown spores from Indonesia (Riley & Jubb, 1999). Parts of eastern Australia, Fiji and Papua New Guinea were reported to be still free from the disease (Braithwaite *et al.*, 2004). The growing significance of this pathogen is clearly evident with the flowing research papers on various aspects of the disease *viz.* host resistance, pathogen variability and diagnosis, management etc., during the past decade. Antony (2008) comprehensively reviewed the status of sugarcane smut in Australia and discussed about the political economy of biosecurity incited due to severity of the disease. More than 70% of Australia's sugarcane varieties were susceptible to smut before 1998. It was not possible to completely eradicate the disease by the time smut was noticed, as it has spread in the whole area under sugarcane cultivation. The immediate management strategy devised was to advocate ploughing out canes with more than 5% infection and switching over to resistant varieties. Since then, smut resistance became an objective of varietal selection in Australia (Croft & Berding, 2005).

3. Epidemiology

A detailed epidemiological study on sugarcane smut was made by Bergamin *et al.*, (1989), who recorded alarming proportions of smut in Brazil. The increase in incidence was found to be associated with varietal susceptibility and increasing age of the crop. The first appearance of the apical whips was found to coincide with around 120 days of planting. The second flush of whip emergence produces an enormous quantity of teliospores and these account for infecting the terminal and lateral buds in the rapidly growing crop. The infected buds may remain dormant and may germinate to produce lateral whips in the third flush of whip production. The infection producing the third level of whips is believed to be critical in the epidemiology of the disease.

4. Pathogen

Germination of smut spores occur on the internodal surface (Fig. 4), which was followed by the formation of appressoria on the inner scales of the young buds and on the base of the emerging leaves. Entry into the bud meristem occurs between 6 and 36 h after the teliospore deposition (Alexander & Ramakrishnan, 1980). Hyphae are found throughout the plant mostly in the parenchymatous cells towards the lower internodes. In the upper internodes, the hyphae are progressively built up culminating in the formation of whip (sori with teliospores). Infective mycelia penetrate through the buds at each node and systemically colonize the apical meristem. Infective buds in mature plants are either symptomatic as whip at the end of stalk or remain asymptomatic hidden in buds up to the next season (Agnihotri, 1990).

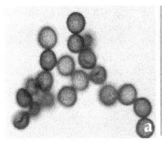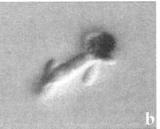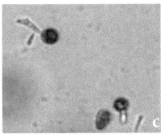

Fig. 4. **Smut teliospores and its germination.** a) Teliospores from whip. b) & c) germination of teliospores.

Sexuality has been demonstrated in the smut pathogen by Alexander and Srinivasan (1966), who showed that it was bipolar, that a combination of two sporidia belonging to opposite sexes was necessary for successful infection and the degree of virulence varied with the combination of haplonts. The existence of physiological specialization has been demonstrated by Alexander & Padmanaban, (1992) and Amire *et al.*, (1982). Classification of races of *U. scitaminea* is based on differences in spore morphology, germination characteristics or pathogenic nature (Sydow,1924). Similar to other species of *Ustilago*, the sugarcane smut fungus is a parasite of young meristematic tissues and gains entry into the host, exclusively through the bud scales (Fawcett, 1942). The pathogen develops systemically throughout the stalk, but teliospores are formed only in peripheral tissues of the whip-like structure. The fungus is capable of mutating and hybridizing in nature in order to produce new virulent pathogenic races (Waller, 1970).

Piepenbring *et al.*, (2002) regrouped the generic position of the sugarcane smut pathogen and renamed it as *Sporisoriun scitamineum*. Three species of smut fungi (Ustilaginales, Basidiomycota) of economic importance, *Ustilago maydis* on corn, *U. scitaminea* on sugar cane, and *U. esculenta* on *Zizania latifolia*, were investigated in order to define their systematic position using morphological characteristics of the sori, ultrastructure of teliospore walls, and molecular data of the LSU rDNA. LSU rDNA analysis suggested that *U. scitaminea* belong to the genus *Sporisorium*. The sugarcane smut fungus develops sori with whip-shaped axes corresponding to columellae and henceforthe, *U. scitaminea* is called *Sporisorium scitamineum*.

5. Variability

Information on the prevalence and distribution pattern of races/pathotypes in *S. scitamineum* in an area is required for effective deployment of host resistance. Schenck (2003) recorded incidence of smut in one variety (H78-7750), considered to be completely resistant in several seed fields on Maui, indicating the possible emergence of a new race of the smut fungus in Hawai. The new smut race was included in breeding program susceptibility screening, keeping in mind, that smut resistant varieties should also be treated and monitored even though the appearance of new smut races was presumed to be quite rare. The use of differential hosts is a viable option for the evaluation of pathogenic variability. However, not much of information on the use of differential hosts is available in sugarcane against the smut pathogen. Gillaspie *et al.*, (1983) used seven sugarcane clones

(*Saccharum* interspecific hybrids) for inoculation with *S. scitamineum* isolates collected from Argentina, Florida, Hawaii, Taiwan, and Zimbabwe. Six different isolates (races) could be differentiated on five of the clones under greenhouse conditions and it was concluded that this method is a valid, rapid method for isolate separation when the correct differential clones are used. It was also observed that the environment effects on the teliospores might be confounded with genetic differences amongst the test isolates which might probably complicate breeding for smut resistance. Smut pathogen being biotrophic, the inoculum henceforth was to be maintained in the standing cane as teliospores, however the fungus has been successfully cultured in an artificial medium in the recent past. Slow growing fluffy white mycelia was observed from actively growing meristem tips cultured under aseptic conditions (Fig. 5), which was further used for molecular characterization of pathogen variability. Smut isolate collection is made from different representative sugarcane growing areas in India and the pathogen variability is being investigated using differential hosts and molecular markers *viz*. RAPD, SSR etc (Ramesh Sundar *et al.*, 2011 - personal communication).

The 20th century saw the steady spread of sugarcane smut to almost all sugar industries of the world (reviewed by Presley, 1978). A widely adapted, stable smut pathotype may have been involved in this spread, explaining the lack of genetic variation in isolates collected from countries outside of Asia. Pathogenic races of sugarcane smut have been observed in several countries including two races (A and B) from Hawaii (Comstock & Heinz, 1977) and three races (1, 2, 3) reported in Taiwan (Leu *et a.l*, 1976). However, Ferreira and Comstock (1989) considered the true prevalence of races to be controversial. Many claims were based on the reaction of the same cultivar in different countries, but the interpretation of these claims was confused by test-to-test variation and the use of different inoculation methods in different countries.

Fig. 5. *In vitro* **culture of smut dikaryotic mycelium.** a) & b) depicts existing morphological variations amongst isolates.

DNA-based markers have been known to detect and measure the variability among individuals and work on molecular characterisation of smut pathogen variability is being carried in many laboratories worldwide in the recent past. Combined application of molecular diagnostic tools along with use of differentials could be an appropriate and reliable approach for studying pathogen variability in *S. scitaminieum*. Braithwaite *et al.*, (2004) employed amplified fragment length polymorphisms (AFLPs) to assess genetic variation between 38 isolates of the sugarcane smut fungus representing 13 countries. The study identified a divergent group of isolates from Southeast Asia. *S. scitamineum* is phenotypically variable with regard to morphology, cultural characteristics and pathogenicity (Abo & Okusanya, 1996). These phenotypic differences appear to be greater than the genetic differences as detected by the neutral AFLP markers, suggesting that these phenotypes correlate with minor changes in the genome or possibly in single genes. They could also be indicative of environmental differences and/or gene expression differences. The results further suggested that alternative fingerprinting techniques, such as simple sequence repeats (SSRs or microsatellites), might provide higher sensitivity and generate more polymorphisms to reveal the existence of yet other clusters.

Xu *et al.*, (2004) studied the genetic diversity of sugarcane smut fungus representing different provinces in Mainland China applying RAPD. Dendrogram of UPGMA cluster analysis revealed that 18 isolates of the fungus were clustered into six groups according to the dissimilarity coefficient of 0.70. The results of cluster analysis suggested that the molecular variation and differentiation could be associated with geographical origin to some extent, but not applicable to all isolates. It might be due to the frequent exchange of sugarcane varieties and clones in the recent years. Molecular diversity analysis observed no relationship between pathogen variability and host origin.

Singh *et al.*, (2005) estimated intraspecies diversity within *Ustilago scitaminea* isolates from South Africa (SA), Reunion Island, Hawaii and Guadeloupe using RAPDs, *b*E mating-type gene detection, rDNA sequence analysis, and spore morphological studies. Mycelial DNA of the South African isolate shared 100% sequence identity with that of mycelial DNA cultured from *in vitro* produced teliospores of the parent cultivar. Overall the ITS1 and ITS2 regions were found to have 96.1% and 96.9% sequence identity with a total of 17 and 21 base changes, respectively, amongst the isolates. The Reunion Island isolate was shown to be most distantly related by 3.6% to the other isolates, indicating a single clonal lineage. The lack of germination in teliospores from Guadeloupe might be attributed to changes in temperature and humidity during transportation.

Raboin *et al.*, (2007) investigated the genetic diversity and structure of different populations of the smut fungus worldwide using microsatellites by subjecting 77 distinct whips (sori) collected in 15 countries worldwide. Results indicated that the genetic diversity of either American or African *S. scitamineum* populations was found to be extremely low and all strains belonged to a single lineage. This lineage was also found in some populations of Asia, where most *S. scitamineum* genetic diversity was detected, suggesting that this fungal species originated from this region. The results obtained in this study thus suggested that the use of resistant cultivars to *S. scitamineum* might be an efficient and durable strategy to control sugarcane smut outside Asia.

Comstock et al., (2007) comprehensively reviewed the status of genetic diversity in S.scitamineum and summarized in line with the results presented during the International Sugarcane Technologists workshop 2006. It was concluded, that the fungus originated in Asia and was disseminated to other continents on rare occasions. It was also indicated that, the resistance reaction of sugarcane clones tested in various countries was strongly influenced by the environment. The possibility of using Near Infra Red spectroscopy (NIR) in prediction of disease resistance rating for smut disease was investigated. The results were promising and the model provided acceptable predicted ratings for all the clones.

Munkacsi *et al.*, (2007) suggested that domestication and cultivation of crop plants did not drive divergence and speciation of smut species on maize, sorghum, and sugarcane. The results obtained greatly weakened a hypothesis, that the speciation of crop pathogens is the necessary result of agricultural practices, and further, showed that these fungi diverged in natural populations of the fungus and host. Most importantly, the findings demonstrated that the domestication process very likely retained symbioses between the crops and scores of microbes, which had co-evolved in ancestral, natural populations. Fattah *et al.*, (2009) attempted genotyping of the races of *Ustilago* species in Egypt using the chitinase gene primers. The study concluded that chitinase genes are the most suitable for genotyping study between sugarcane smut fungal isolates. The results obtained by differential display techniques showed that there were at least 10 different races from the *Ustilago sp.* in Egyptian field. Nzioki *et al.*, (2010) attempted to identify presence of physiological races of sugarcane smut and the results suggested possible existence of smut races in Kenya.

6. Diagnosis

Correct diagnosis of pathogens is the primary requirement in any sound disease management practice. It is important for the identification of pathogens, breeding crops for resistance to pathogens and epidemiological studies. Conventional approaches involve use of microscopy combined with specific stains for histopathological studies. Serology-based diagnostic techniques proved to be equally efficient in the diagnosis of sugarcane pathogens. Sinha and Singh (1982) developed a staining technique using trypan blue for the detection of smut hyphae in nodal buds of sugarcane (Fig. 6). This rapid staining technique enabled detection of hyphae of *S. scitamineum* in the growing points of nodal buds of sugarcane. The results concluded that this whole detection process can be completed within

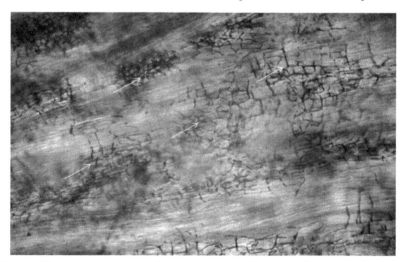

Fig. 6. **Trypan blue staining of smut fungus.** Arrows indicate proliferation of inter and intracellular mycelial growth of smut fungus in nodal buds of sugarcane

4 hour period and smut infection can be detected in buds earlier than the symptom expression on planting. This technique finds application in quarantine and seed certifying agencies for screening sugarcane seed material. Nallathambi *et al.* (1998) observed that the trypan blue staining also detected smut pathogen colonization in some clones, which escaped infection in the field. Further this staining technique was found to be very rapid, precise and allows a large number of samples to be tested in a short period. An indirect ELISA technique was standardized for screening large number of sugarcane clones for smut pathogen detection (Nallathambi *et al.*, 2001). This work outlines smut antigen preparation and its appropriate dilutions for an early detection of smut pathogen at symptomless infection stage in sugarcane settlings. Acevedo and Pinon (1996) developed an indirect immunofluorescence technique for the diagnosis of *S. scitamineum* infection in sugarcane. Optimization of the methodology resulted in the best dilution of the antiserum for efficient detection of the smut pathogen.

Technological advances in PCR-based methods, such as real-time PCR, allow fast, accurate detection and quantification of plant pathogens and are now being applied to practical problems. Albert and Schenck (1996) successfully amplified *S. scitamineum* with the use of primers based on the *U. maydis b*E mating type gene. Sequence analysis of the PCR amplicon yielded around 70% homology with the *b*E in *U. maydis* and *U. hordei*. The PCR-product of 459 bp is specific to *S. scitamineum* and it has been validated successfully by many researchers. Singh *et al.* (2004) demonstrated that PCR assay was extremely sensitive in detecting the presence of the pathogen and yielded a positive response in plantlets inoculated with sporidia and observed that PCR assay was significantly better for smut detection than microscopy. Whilst the PCR assay and microscopy may be used to detect the smut pathogen in plantlets not exhibiting symptoms of infection, it was concluded that there was no relationship between the presence of the pathogen and plant resistance. Yudilay *et al.*, (2004) critically evaluated different diagnostic methods *viz.* conventional, optic microscopy, serological and molecular of the sugarcane smut *Ustilago scitaminea* Syd and weighed out the advantages and disadvantages of each one of them according to sensibility, efficiency and possibilities. Jorf and Izadi (2007) isolated and purified yeasts-like and dikaryotic mycelial colonies of the sugarcane smut pathogen and concluded that PCR assay and microscopic study could be used effectively to detect the presence of smut pathogen in settlings not exhibiting symptoms of infection. The results of an investigation also revealed that PCR assay resulted in more early detection of the pathogen (Ramesh Sundar *et al.*, 2011 – Personal communication).

7. Host resistance

7.1 Screening for smut resistance and its biochemical indices

Releasing disease resistant varieties has been the prime management strategy to reduce the yield loss caused by the fungal pathogens in sugarcane. Burner (1993) evaluated the smut resistance of *Saccharum* spp. *viz.* *S. officinarum, S. barberi, S. sinense, S. robustum, S. spontaneum, Erianthus* spp. section Ripidium, and *Saccharum* interspecific hybrids (cultivars). The study revealed that clones of *Erianthus* spp. section Ripidium were the most resistant clones and clones of *S. officinarum* and *S. robustum* were the most susceptible amongst the six taxonomic groups studied. Clones from India seem to have moderate levels of resistance, whereas those from Indonesia and Philippines were found to pick more than 50% infection on screening. Glenn *et al.*, (1998) reported that *Erianthus* spp. and other wild relatives of *S. officinarum* are being used in an intergeneric sugarcane breeding programme in an effort to increase sugarcane resistance to sugarcane mosaic virus and sugarcane smut (Burner *et al.*, 1993). However, breeding for disease resistance has been complicated by the frequent emergence of new pathogenic variants, which overpower the resistant varieties, as witnessed from the withdrawal of erstwhile ruling varieties from commercial cultivation. Even now, the benefit from such varieties could not be harnessed to maximize the sugarcane productivity in many of the developing countries including India, by virtue of its extreme susceptibility to important diseases like red rot and smut.

The evaluation of varieties for smut resistance is generally similar throughout the world. The rating is done based on the percentage of infected stools. Most countries employ the 0-9 disease scale of Hutchinson (1970), but differ in their assignment of infection percentage to disease rating. In evaluating smut resistance, due considerations are to be given for the percentage of infection. Waller (1970) made a pioneering work in comparing different methods of smut inoculation. Injection inoculation may induce greater smut infection than dip inoculation and the results indicated that cultivars could respond differently to the two methods of inoculation. Screening for smut reaction typically involves a dip inoculation assay in which nodal buds are immersed briefly in a suspension of teliospores, and then planted in a greenhouse. The periodical observation of smut incidence is recorded and on the basis of cumulative final percentage of disease incidence, varieties are graded as R, MR, MS S, and HS (Alexander & Padmanaban, 1988). It is a means of pre-screening large numbers of new sugarcane genotypes for resistance to smut disease. Highly resistant and resistant cultivars by this assay can then be field tested for validation and verification. Susceptible genotypes on the other hand, can be detected early and discarded. Evaluation can take place in a greenhouse or in the field (Alexander & Padmanaban., 1992).

Singh *et al.*, (2005) screened tissue-cultured plantlets of three sugarcane (*Saccharum* spp.) cultivars having a known field smut reaction for smut susceptibility and established corroboration of the *in vitro* results with that of screening under field conditions. Olweny *et al.*, (2008) critically evaluated the smut inoculation techniques in sugarcane seedlings and explored the possibility of screening for smut resistance at the seedling stage. Wound paste method recorded the highest incidence of smut whip production, followed by paste, however, soaking method had the lowest incidence of smut.

Basically sugarcane smut resistance mechanism is characterized into bud resistance (infection resistance) and inner tissue resistance (colonization resistance) (Dean, 1982). It was observed by Singh and Budharaja (1964), that hyphae will not penetrate cells of the scale leaves. Hence buds tightly enclosed with the scale leaves have a better chance of escaping infection. On this basis, Waller (1970) hypothesized that varietal resistance was determined by bud morphological characteristics. Structural characterization of sugarcane buds could provide clues for classification of test clones according to its smut resistance. da Gloria *et al.*, (1995) established an association between the bud structural characteristics and the cultivar resistance. Presence of outer most scales were hypothesised to provide protection against the bud invasion of the smut pathogen.

It is well documented that plants have evolved effective resistance mechanisms, that enable them to defend themselves against pathogen attack. Many reports are available on this front involving sugarcane and the smut pathogen, which suggested a chemical resistance mechanism than the morphological one. Glycosidic substances isolated from fresh bud scales were found to have linear association with smut resistance. Lyold and Pillay (1980) identified flavonoids as inhibitors of teliospore germination and established a relationship between smut sensitivity to smut and polyamine conjugation. Infection of buds from both sensitive and resistant cultivars of sugarcane with teliospores of S. scitamineum lead to remarkable increase of both free and conjugated polyamines. Conjugation of polyamines to phenolics has often been described as a defence mechanism against infection of several higher plants by viruses and fungi. Conjugation mainly affects tyramine conjugated to ferulic and hydroxycinnamic acids (Fleurence & Negrel, 1989) or spermidine and spermine conjugated to hydroxycinnamic acids (Hedberg et al., 1996). Lloyd and Naidoo (1983) proposed that phenolics are produced as a linear response of resistance acquisition against smut infection, it can be hypothesized that the conjugation of polyamines to phenolics can nullify the microbicidal action of these compounds. A negative relationship between glycosidic substance content in bud scale and resistance of sugarcane varieties to smut was observed, indicating that the glycosidic substance in bud scale might be a chemical mechanism of resistance against the infection of Ustilago scitaminea.

The level of different polyamines and the possible conjugation to phenolics in mature organs of S. scitamineum-infected and non-infected sugarcane plants has been suggested to be correlated with smut susceptibility, indicating that polyamine conjugation to phenolics may act as a mechanism of resistance or defense against this disease. Legaz et al. (1998) attempted to study the relationship between the sensitivity of resistance to smut with the accumulation of free or conjugated polyamines in sugarcane tissues, and observed that infectivity and development of fungal mycelium in sensitive buds could be clearly correlated with a dramatic increase of both SH and PH-spermidine and spermine.

Rodriguez et al. (2001) hypothesized key role of the oxidative burst on the early sugarcane response against the S. scitamineum infection. Results suggested that ethylene could be inducing sugarcane transcripts related to auxins and defense proteins. Xu et al. (1994) reported that infection by Ustilago scitaminea resulted in increase in peroxidase (POD) and invertase activity in both resistant and susceptible sugarcane plants. The results suggested that POD activity could be used as an index for smut resistance in sugarcane. PAL, TAL, CoA-ligase specific activities and chlorogenic acid, total flavone contents were measured in sugarcane varieties with different resistance to smut after inoculation with Ustilago scitaminea Syd.. PAL, TAL, CoA-ligase activities of highly resistant varieties were higher and maintained longer time than those of highly susceptible ones. At the same time, the accumulation of chlorogenic acid, total flavone contents in highly resistant varieties was not only earlier, but also quantitatively higher. Therefore, the results suggested that strengthening of phenylpropanoid metabolism induced by Ustilago scitaminea might be an important aspect of sugarcane post-infectional resistant mechanism to smut. The increase of activities of POD and acid invertase was also observed in leaves of sugarcane infected by sugarcane chlorotic streak virus (Wang et al., 1995). Singh et al., (2002) observed an increase in the ascorbic acid content in leaf, bud, apical meristem, lateral shoots as well as in juice of smut affected stalks in two smut susceptible varieties. It was presumed that the enhancement in the ascorbic content in smut affected stalks might be due to the production of ascorbic acid accelerating enzymes by the pathogen or by the interaction of host-parasite.

The role of sugarcane glycoproteins in the resistance of sugarcane to smut was examined by many researchers. Sugarcane produces two different pools of glycoproteins containing a heterofructan as glycidic moiety and tentatively described as high molecular mass (HMMG) and mid-molecular mass (MMMG) glycoproteins (Legaz *et al.* 2005). Analysis of both HMMG and MMMG by capillary electrophoresis revealed that MMMG fraction contains two cationic and four anionic components, whereas only one cationic and four anionic proteins are separated from the HMMG fraction (Legaz *et al.* 1998). These glycoproteins affected polarization of the cytoplasm during spore germination, impaired germ tube protrusion and germination of the spores ultimately. These could be considered as factors contributing to smut resistance (Martınez *et al.* 2000). As their amount increases after infection with smut teliospores in resistant, but decreases in susceptible varieties after infection with smut teliospores. Fontaniella *et al.*, (2002) ascertained the role of these glycoproteins in sugarcane smut resistance and recorded that Methyl jasmonate did not produce an elicitation response for glycoprotein synthesis in sugarcane. On the contrary, salicylic acid, secreted by germinating spores of *S. scitamineum* acted as an elicitor of glycoprotein production, and the elicitation process could be experimentally simulated by using this compound instead of spore inoculation. However, the quantitative response of sugarcane stalks to the infection in order to produce defence glycoproteins is higher than that obtained by infiltration of salicylic acid in plant tissues. The results opened up the possibility of the secretion of a co-elicitor, other than salicylic acid and unidentified as yet, seems to be required for the complete response. It has been proposed that the inhibition of teliospore germination constitutes a defence mechanism involved in the general pattern of the resistance of sugarcane to the smut pathogen.

Millanes *et al.* (2005) examined the role of sugarcane glycoproteins in regulating the cell polarity of *S. scitamineum*. Smut teliospores were found to be able to change the pattern of glycoprotein production by sugarcane, thereby promoting the synthesis of different glycoproteins that activate polarization after binding to their cell wall ligand. The study further demonstrated that smut teliospores were able to change the metabolism of parenchymatous cells of resistant sugarcane cultivars by increasing glycoprotein production. The results proposed that inhibition of teliospore germination constitutes a defense mechanism involved in resistance of sugarcane to smut. Millanes *et al.*, (2008) hypothesized that the inhibition of smut teliospores germination by sugarcane glycoproteins, HMMG and MMMG, could be specifically related to actin polymerization. High molecular mass elicitors (proteins or glycoproteins) were previously detected in *Colletotrichum falcatum* (Went) (Ramesh Sundar *et al.*, 2002), but these types of compounds from smut mycelium did not show biological activity.

Inoculation with the smut pathogen produced new phenolics, that increased the level of Hydroxy cinnamic acids (HCA) and their derivatives to enhance the synthesis of lignin and strengthening of the cell wall in the sugarcane cultivar resistant to *S. scitamineum*. de Armas *et al.*, (2007) observed that the sensitivity or resistance of sugarcane to smut can be related to changes in the levels of free phenolic compounds, and phenylalanine ammonia-lyase (PAL) and peroxidase (POD) activities in the leaves. Elicitors from *S. scitamineum* enhanced the activity of PAL and consequently increased the levels of hydroxycinnamic and hydroxybenzoic acids. However, a decrease in the amount of free hydroxycinnamic acids was found, when the highest PAL activity was reached. It was concluded that monitoring changes in leaf phenolic compound concentrations, PAL and POD activities in response to soluble elicitors extracted from *S. scitamineum* mycelium could afford reliable analyses of the resistance of sugarcane to smut. A resistant cultivar needs to maintain a high level of PAL activity without accumulation of free hydroxycinnamic acids. Increase in POD activity is important in the defence mechanism, but it is not a determinant for the defence mechanism. This model was proposed for the screening of smut resistance levels of different sugarcane cultivars and it would help breeding programs to characterise promising clones. Further it was concluded, that it is possible to say that the metabolism of phenylpropanoids seems to be directly related with resistance to smut.

Santiago *et al.*, (2008) identified smut-elicitor fractions as resolved by capillary electrophoresis. Those inducing the highest biological activity corresponded to negatively charged proteins, peptides or glycopeptides of medium molecular mass. These compounds enhanced the accumulation of free phenolics, mainly hydroxycinamic acids, by activation of PAL in the resistant cultivar, and hydroxybenzoic acids in the susceptible cultivar. Another important difference in the resistant cultivar was the enhancement of POD-an enzyme that uses free phenolics as substrates for the activation of important mechanisms of resistance of sugarcane leaves to the fungal pathogen. Santiago *et al.*, (2010) further correlated changes in the levels of phenolics substances, induced by a smut elicitor, which resulted in increase in thickness of the lignified cell walls and thus could contribute as a possible mechanical defense response to the potential entry of the smut pathogen. It was hypothesized that lignin deposition in supporting tissues might be indicative for biochemical and structural resistance responses in sugarcane.

7.2 Molecular markers for smut resistance

In order to understand the mechanism behind disease resistance in sugarcane, recent studies include molecular approaches involving Genomics and Proteomics tools. With the advent of such sophisticated tools of biotechnology, it has now become possible to gain better understanding on sugarcane-pathogen interaction. The processes that determine the outcome of an interaction between a microbial pathogen and a host plant are complex. Understanding the molecular details of these interactions, such as the pathogen genes required for infection, effective host defense responses and mechanisms by which host and pathogen signaling networks are regulated, might be utilized to design new plant protection strategies. A major limitation, however, is the poor availability of genetic tools in sugarcane because of the genomic complexity due to its polyploidy nature. Nevertheless, further characterization and functional analysis of the genes that are identified in the Sugarcane EST (SUCEST) program can lead to a more comprehensive understanding of sugarcane-pathogen interactions.

AFLP-based genetic mapping strategy by Raboin *et al.*, (2001) focussed on a cross between cultivar R 570 (resistant) and cultivar MQ 76/53 (highly susceptible), which showed a segregation for smut resistance in a preliminary field trial. The findings established correlations between segregating markers and resistance to smut and discussed the possibility of identifying the different components involved in smut resistance and the interest of locus specific markers (SSR, resistance gene analogs, etc) to refine the genetic map. Thoakoane and Rutherford (2001) explored the possibility of isolating differentially expressed genes in sugarcane in response to challenge with the smut pathogen by using cDNA -AFLP. Sequence homology searches of isolated genic fragments have identified a putative chitin receptor kinase, a Pto ser/thr protein kinase interactor, and an active gypsy type LTR retro-transposon expressed differentially in the resistant variety in response to challenge. Sugarcane genes encoding proteins homologous to chitinases, as well as transcripts related to the pathways of both phenylpropanoids and flavonoids were shown to be involved in the sugarcane resistance after 7 days of *S. scitamineum* infection.

Sequence analysis of genes differentially expressed in response to challenge by smut has identified putative receptors involved in the signalling of resistance mechanisms, transcription factors, and enzymes involved in phenylpropanoid-flavonoid metabolism (Heinze *et al.*, 2001). Two full-length thaumatin (PR5) antifungal protein coding sequences have been isolated and are available for use as transgenes. Constitutive expression of acidic thaumatin suggested the involvement of SA signalling in sugarcane buds, as does the presence of a putative SA inducible cell-wall bound receptor kinase.

Genes encoding NBS-LRR-like proteins, protein kinases, and proteins related to both auxin and ethylene pathways were found to contribute to stable resistance against the sugarcane smut pathogen (Borras *et al.* 2005). The studies by Butterfield *et al.*, (2004) and Hidalgo *et al.*, (2005) demonstrated that subtractive or differential display techniques could be used to identify genes, that are activated during biotic stress responses, such as those induced by pathogens, and allow the isolation of rare transcripts elicited as part of the plant's resistance response. Results of the Northern blot analysis indicated that mRNA levels of genes, that are homologous to four of those transcript-derived fragments (TDFs) were highly induced in resistant somaclones inoculated with *S. scitamineum*, while no or low expression was observed in the susceptible parental lines, thus confirming the differential expression pattern. The differential expression of a number of sugarcane genes upon inoculation with the sugarcane smut fungus *S. scitamineum* was affiliated with disease resistance, as it makes sense that they should have the potential to be developed into markers for resistance.

In sugarcane, the expression pattern of a putative ethylene receptor (SCER1) and two putative ERF transcription factors (SCERF1 and SCERF2) showed differential responses to interactions with pathogenic and beneficial microorganisms, which suggested that they might participate in specific ethylene signaling cascade(s), that can identify a beneficial or pathogenic interaction (Cavalcante *et al.* 2007). Que *et al.*, (2008) attempted to isolate resistance gene analogs (RGAs) from sugarcane (*Saccharum officinarum* Roxb.) with primers targeting the conservative sequences of nucleotide-binding site (NBS). A full-length cDNA of cRGA1 (Accession number: EF155648), termed *SNLR* gene, was cloned and its expression profile under the treatment of *S. scitamineum*, SA and H_2O_2 was investigated by real-time RT-PCR (Accession number: EF155654). The results showed that *SNLR* gene could be to some extent influenced by *S. scitamineum* and SA, but not by H_2O_2. Based on the results of Que *et al.*, (2008), it was hypothesized, that this might be due to the reason that the *NLR* gene does not occur *via.* an H_2O_2 dependent pathway or involves a different mechanism. Further work on the functional genomics part involving transgenic complementation, gene knock-out or other experiments would add more information to establish its function in smut resistance. Subsequent investigations by Que *et al.*, (2009) indicated the presence of non-TIR-NBS-LRR type resistance genes only in the genome of sugarcane. The 11 RGAs, together with *RPS2* and *Xa1*, were clustered into one group, and *N* and *L6* were in another group. One RGA, termed *PIC* (EF059974), was validated through real-time PCR. The result showed that the expression of *PIC* gene was induced by *S. scitamineum* and salicylic acid, but inhibited by hydrogen peroxide. The *PIC* gene had constitutive expressions in leaves, stalks, and roots of sugarcane, with the strongest expression in leaves, which has a proven correlation with resistance to several diseases in sugarcane.

Lao *et al.*, (2008) established the involvement of major plant signaling pathways during the first 72 h of interactions between sugarcane and *S. scitamineum*. A differential expression study on the *Saccharum* spp.–*S. scitamineum* pathogenic interaction was undertaken involving a susceptible (Ja60-5) and a resistant (M31/45) genotypes. A total of 64 transcript-derived fragments (TDFs) were found to be differentially expressed by using cDNA-AFLP analysis, wherein a majority (67.2%) of the differential TDFs was found to be up-regulated in the resistant M31/45 cultivar. The plant response against *S. scitamineum* infection was complex; representing major genes involved in oxidative burst, defensive response, ethylene and auxins pathways during the first 72 h post-inoculation. Results of this study suggested that the genes involved in the oxidative burst and the lignin pathways are vital for the initial sugarcane defense against the *S. scitamineum* infection. Segregation studies of the differentially expressed genes in "R" and "S" sugarcane progenies may provide more insight into the genetic basis of smut resistance in sugarcane.

8. Quarantine

In Australia, Sugarcane smut was identified as a high-risk exotic disease in a pest risk analysis conducted, and a contingency plan to deal with incursions was prepared in 1997, since its first time report in Australia in July 1998. Quarantine regulations were enacted in Queensland and New South Wales to reduce the risk of spread by plant material or appliances. The ORIA cane growers cooperated by ploughing out heavily infested fields and had removed all susceptible cultivars by 2001. This has reduced the risk of wind-borne spread from the ORIA. Nearly 20% of the germplasm collections maintained at Thailand recorded smut incidence (Jaroenthai *et al.*, (2007), which has resulted in the reduction of yield, CCS, and brix by 8–18%, 7–13% and 17–43%, respectively. Infection and severity of the smut disease normally increased in ratoon cane, because smut spores can spread with wind, rain and the pathogen can survive in dry soil for 2–3 months. However, level of infection and severity also depended on the resistance of each variety.

Magarey *et al.*, (2008) highlighted the perceived threats due to diseases and insect pests to *Saccharum* germplasm in Australia and neighbouring countries. An Australian centre for International Agricultural Research (ACIAR) funded program on conservation of germplasm was implemented in Papua New Guinea, Indonesia and Northern Australia. Since these areas constitute the centre of diversity for various *Saccharum* spp. there was increasing threats to the germplasm observed. Smut was perceived as one of the possible threats in Australia, as there was regular exchange of germplasm from neighbouring countries. In view of the alarming situation, a concerted breeding program was initiated, in which more than 1500 Australian clones have been screened for smut resistance in Indonesia.

9. Management

Seed selection and selective rouging of infected clumps would assure a healthy crop. Periodical observations of the standing crop and removing the whips would considerably reduce the amount of pathogen inoculum, thus preventing further build-up of the pathogen. Studies reported that smut teliospores lack dormancy and hence could not survive in soil or debris in the absence of buds. This prompted advocation of deep ploughing and irrigating the fields, which will allow germination of the teliospores and would eventually die off in the absence of buds.

9.1 Physical control

Various hot water treatments have been reported to be effective in controlling the smut pathogen residing in the planting setts. The loss of bud germination due to inappropriate temperature settings needs to be handled properly (Srinivasan & Rao, 1968). Hardening of setts prior to hot water treatment was observed to considerably improve upon the germination of the buds. The efficacy of moist hot air treatments have been reported by Misra *et al.*, (1978). Gupta *et al.*, (1978) reported production of thicker and heavier canes with an increased number of millable canes due to hot water treatment.

9.2 Chemical treatment

Vangaurd and Bayleton treatment inhibited smut development from systemically infected seed pieces (Comstock *et al.*, 1983). It is a recommended practise to subject the planting setts to a hot water treatment @ 52°C for 30 min combined with a chemotherapy using 0.1% Triademiphon - Bayleton (Mameghmay, 1984). This treatment was found to completely eliminate the sett-borne infection of smut.

Wada *et al*, (1999) suggested effective strategies for the management of sugarcane smut, *viz.* pre-plant heat therapy of planting setts; pre-plant fungicidal dips of planting setts and screening of sugarcane clones for identification of resistant varieties. It was observed that these single strategy controls might not be adequate for many sugarcane pests and diseases including smut, thus opting for IPM strategy, which would be a viable and successful smut management strategy. The need for continuing tests of different fungicides with varying modes of action for smut control has been discussed by Wada (2003). The best disease control was obtained with pyroquilon at 4.0, carbendazim+maneb at 4.57 and metalaxyl+carboxin+furathiocarb at 9.9 g a.i. Kg^{-1}, respectively. The efficacy of pyroquilon and metalaxyl+carboxin+furathiocarb, which *hitherto* were used as a seed treatment in cereals, revealed the availability of alternative uses for them in smut control.

Joyce *et al.*, (2008) attempted to utilise smut resistant varieties in genetic modification research programs leading to commercial GM crop development in Australia. Protocol optimization was done for selecting an efficient tissue culture medium to produce embryogenic calli with high transformation efficiency.

10. Conclusion and future perspective

Sugarcane smut continues to be a serious threat to sugarcane production in different countries. Integrated disease management strategy is the viable option in smut disease control, rather than resorting to a single method. Recommended phytosanitary practices like seed selection, roguing of infected clumps etc is the best possible way to reduce smut inocula levels. Research on identifying sources of smut resistance in the germplasm and progenies needs to strengthened. Development of smut resistant varieties to the current pathotype of *S. scitamineum* with a focussed breeding program, combined with clean cultivation practices would lead to successful management of smut disease in sugarcane. An understanding of the existing race picture of the pathogen is a pre-requisite for disease management, which could be accomplished by harnessing the tools of biotechnology. Recent literature attempts at throwing more light on understanding the biochemical and molecular basis of smut resistance in sugarcane. Though limited information is available regarding the sources of resistance, molecular tools are now available to identify suitable markers that can be relied upon for supporting the conventional breeding approaches. Similarly molecular diagnostic tools should be developed for a rapid and precise detection of the smut pathogen in seed cane. This supplemented with a strict quarantine regulations would prevent introduction of the disease into a new region and ensure supply of disease free seed material for planting. Information availability on the epidemiology of smut disease is very limited and more emphasis should be given to study the influence of critical weather parameters on smut severity, as this would lead to a better understanding on the impact of climate change on this important disease of sugarcane. Also efficient decision-support systems need to be developed for smut disease forecast, thus will result in the development of precise forewarning systems of a possible outbreak of the disease. In addition to the existing control measures, novel strategies should be thought of to explore the possibility of inducing systemic resistance against the smut pathogen. Further with the identification of candidate defense genes, development of transgenic sugarcane with built-in resistance to smut is to be looked into for the future.

11. Acknowledgements

The authors thank Dr. N. Vijayan Nair, Director of the Sugarcane Breeding Institute (ICAR), INDIA, for providing facilities and continuous encouragement. The authors greatly appreciate and duly acknowledge the excellent ongoing work on Sugarcane smut being performed by the research scholars of the Pathology group of the Institute.

12. References

Abo, M.E. & Okusanya, B.A. (1996). Incidence and variability reaction of sugarcane smut (*Ustilago scitaminea* Syd.) isolates in greenhouse and laboratory tests in Nigeria. *Discovery and Innovation*, Vol. 8, pp. 227–231

Acevedo, R. & Piñón,D. (1996). Indirect immunofluoresence for sugarcane smut diagnosis. *Revista Iberoamericana de Micología*, Vol. 13, No. 1, pp. 8-9

Agnitori, V.P. (1990). *Diseases of Sugarcane and Sugar Beet*. Oxford and IBH Publishing Co. Pvt. Ltd., Oxford.

Akalach, M. (1994). First report of sugarcane smut in Morocco. *Plant Disease* Vol. 78, No.5, pp. 529

Albert, H.H. & Schenck, S. (1996). PCR amplification from a homolog of the bE mating-type gene as a sensitive assay for the presence of *Ustilago scitaminea* DNA, *Plant Dis.*, Vol. 80 pp. 1189–1192

Alexander, K. C. & Padmanaban, P. (1992). Smut of sugarcane, In: *Plant diseases of international importance, Diseases of sugar, forest, and plantation crops*. A.N.

Mukhopadhyay, J. Kumar, H.S. Chaube and U.S. Singh. Englewood Cliffs,USA, Prentice Hall: Vol. 4, pp. 1626

Alexander, K. C. & Padmanaban, P. (1988). Smut disease of sugarcane, In: *Perspectives in mycology and plant pathology*, pp. 123-137, Malhotra Publishing House, New Delhi India

Alexander, K.C. & Ramakrishnan, K. (1980). Infection of the bud, establishment in the host and production of whips in sugarcane (*Ustilago scitaminea*). *Proc Int Soc Sug Technol* Vol.17, pp. 1453-1455

Alexander, K.C. & Srinivasan, K. V. (1966). Sexuality in Ustilago scitaminea Syd. *Curr. Sci.* Vol. 35, No.23, pp. 603-604

Amire,O.A.; Trione, E.J. & Schmitt, R.A. (1982). Characterization of pathogenic races of the sugarcane smut fungus by neutron activation analysis. *J. Radioanal. Chem.,* Vol.75, pp. 195-203

Antoine, R. (1961). *Smut.* Sugarcane diseases of the world. J.P. Martin, E.V. Abbott & C.G. Hughes. Elsevier, pp.327 – 354, Amsterdam

Antony, G. (2008). Sugarcane smut: the political economy of biosecurity, Proceedings of the 52nd annual conference of the Australian Agricultural and Resource Econonomics Society Canberra, February 2008

Banihashemi, Z. (1995). The occurrence of sugarcane smut in Mazandarn Province. *Iranian Journal of Plant Pathology* Vol.31, No.1/4, pp. 40 - 41

Bergamin,A.; Amorim,L., Cardoso, C.O.N., Da Silva,W.M., Sanguino, A., Ricci, A. & Coelho, J.A. (1989). Epidemiology of sugarcane smut in Brazil. *Sugarcane*, pp. 211-16.

Borras, O.; Thomma, B.P.H.J., Carmona, E., Borroto, C.J., Pujol, M., Arencibia, A. & Lopez, J. (2005). Identification of sugarcane genes induced in disease-resistant somaclones upon inoculation with *Ustilago scitaminea* or *Bipolaris sacchari*. *Plant Physiol Biochem* Vol. 43, pp. 1115–1121

Braithwaite, K.S.; Bakkeren, G., Croft, B.J. and Brumbley, S.M. (2004). Genetic variation in a worldwide collection of The sugarcane smut fungus *Ustilago scitaminea, Proc. Aust. Soc. Sugar Cane Technol.,* Vol. 26

Burner, D.M, Grisham, M.P. & Legendre, B.L. (1993). Resistance of sugarcane relatives injected with *Ustilago scitaminea. Plant Dis.* Vol. 77, pp. 1221 – 1223

Butterfield, M.K.; Rutherford, R.S., Carson, D.L. & Huckett, B.I. (2004). Application of gene discovery to varietal improvement in sugarcane, *South Afri. J. Bot.* Vol. 70 pp. 167–172

Cavalcante, J.J.; Vargas, C., Nogueira, E.M., Vinagre, F., Schwarcz, K., Baldani, J.I., Ferreira, P.C. & Hemerly, A.S. (2007). Members of the ethylene signaling pathway are regulated in sugarcane during the association with nitrogen-fixing endophytic bacteria. *J Exp Bot,* Vol. 58, pp. 673–686

Comstock, J. C. Croft, B.J., Rao, G.P., Saumtally,S. & Victoria, J.I. (2007). A Review of the 2006 International Society Of Sugar Cane Technologists. Pathology Workshop., *Proceedings of the Int. Soc. Sugar Cane Technol.,* Vol. 26, 2007

Comstock, J.C. & Heinz, D.J. (1977). A new race of culmicolous smut of sugarcane in Hawaii. *Sugarcane Pathologists' Newsletter* Vol. 19, pp. 24-25

Comstock, J.C.; Ferreira, S.A. & Tew, T. (1983). Hawaii's approach to control of sugarcane smut. *Plant Dis.,* Vol. 67, pp. 452-457

Croft, B. & Berding, N. (2005). *Breeding New Smut Resistant Varieties*. BSES Bulletin No 8. BSES Ltd, Brisbane

da Gloria, B.A.; Albernas, M.C. & Amorim, L. (1995). Structural characteristics of buds of sugarcane cultivars with different levels for resistance in smut. *Journal of Plant Diseases and Protection*, Vol. 102, No. 5, pp. 502-508

de Armas, R.; SantiagoB R., LegazB, M.E. & Vicente, C. (2007). Levels of phenolic compounds and enzyme activity can be used to screen for resistance of sugarcane to smut (*Ustilago scitaminea*). *Australasian Plant Pathology*, Vol. 36, pp. 32–38

Dean, J.L. (1982). The effect of wounding and high pressure spray inoculation on the smut reaction of sugarcane clones. *Phytopathology*, Vol. 71, pp. 1023

Fattah, A.I.; Alamri, S. R., Abou-Shanab, A.I. & Hafez, E.E. (2009). Fingerprinting of *Ustilago Scitaminea* (Sydow) in Egypt Using Differential Display Technique: Chitinase Gene the Main Marker. *Research Journal of Agriculture and Biological Sciences*, Vol. 5, No. 5, pp. 674-679

Fawcett, G.L. (1942). Circular, *Estacion Experimental Agricola*, pp. 114, Tucuman

Ferreira, S.A. & Comstock, J.C. (1989). Smut. In: *Diseases of Sugarcane*, Ricaud, C., Egan, B.T., Gillaspie, A.G., Hughes, C.G., pp. 211–229, Elsevier, Amsterdam

Fleurence, J. & Negrel, J. (1989). Partial purification of tyramine feruloyl transferase from TMV inoculated tobacco leaves. *Phytochemistry*, Vol. 28, pp. 733-736

Fontaniella, B.; Márquez, A., Rodríguez, C.W., Piñón, D., Solas, M.T., Vicente, C. & Legaz, M.E. (2002). A role for sugarcane glycoproteins in the resistance of sugarcane to *Sporisorium scitaminea*. *Plant Physiol. Biochem.*, Vol.40, pp. 881–889

Gillaspie, Jr. A.G.; Mock, R.G. & Dean, J.L. (1983). Differentiation of *Ustilago scitaminea* isolates in greenhouse tests. *Plant Dis.*, Vol. 67, pp. 373–375

Glenn, A. E.; Rykard, D.M., Bacon, C.W. & Hanlin, R.T. (1998). Molecular characterization of *Myriogenospora atramentosa* and its occurrence on some new hosts. *Mycol. Res.* Vol. 102, No.4, pp. 483-490

Gupta, S.C.; Verma, K.P., Singh, M.P. & Misra,S.C. (1978). Control of diseases by hot water treatment of sugarcane seed material. *Indian Sug. Crops J.*, Vol. 5, pp. 28

Hedberg, C.; Hesse, M., & Werner, C., (1996). Spermine and Spermidine hydroxy cinnamoyl transferases in *Aphelandra tetragona*. *Plant Science*, Vol. 113, pp. 149-156

Heinz, D.J. (1987). *Sugarcane improvement through breeding*. Elsevier Publications, pp. 455-502

Heinze, B.S.; Thokoane, L.N., Williams, N.J., Barnes J.M. & Rutherford, R.S.(2001). The Smut-Sugarcane interaction as a model system for the integration of marker discovery and gene isolation. *Procedings of the S Afr Sug Technol Ass.*, Vol. 75, pp. 88-93.

Hidalgo, O.B.; Thomma, B.P.H.J., Carmona, E., Borroto, C.J., Pujol M., Arencibia, A. & Lopez, J. (2005). Identification of sugarcane genes induced in disease-resistant somaclones upon inoculation with *Ustilago scitaminea* or *Bipolaris sacchari*. *Plant Physiology and Biochemistry*, Vol. 43, pp. 1115–1121

Huang, S. (2004). Progress of Sugarcane Disease Research in China: Recent Developments. *Sugar Tech.*, Vol. 6, No. 4, pp. 261-265

Hutchinson, P.B. (1970). A standardized rating system for recording varietal resistance to sugarcane disease. *Sugarcane Pathol.* Newsletter, Vol. 5, pp. 7

Jaroenthai, K.; Dongchan, S., Anusonpornpurm, S. & Pliansinchai, U.(2007). Occurrence of sugarcane diseases in the germplasm collection at Mitr Phol Sugarcane research centre at Chaiyaphum, Thailand. *Sugar Cane Technol.*, Vol. 26, pp. 1040-1045

Jorf, A.S. & Izadi, M.B. (2007). *In vitro* detection of yeast like and mycelial colonies of *Sporisorium scitaminea* in Tissue cultured plantlets of Sugarcane using Polymerase chain reaction. *Journal of Applied Sciences*, Vol. 7, No. 23, pp. 3768-3773

Joyce,P.; Brumbley, J., Wang, L.F. & Lakshmanan P.(2008). Tissue culture and biolistic transformation of new smut resistant sugarcane varieties. *Sugarcane International*, Vol. 26, No. 61, pp. 10 -14

Lao, M.; Arencibia, A.D., Carmona, E.R., Acevedo, R., Rodrıguez, E., Leon, O. & Santana, I. (2008). Differential expression analysis by cDNA-AFLP of *Saccharum* spp. after inoculation with the host pathogen *Sporisorium scitamineum*, *Plant Cell Rep.*, Vol. 27, pp. 1103–1111

Legaz, M.E.; de Armas, R., Millanes, A.M., Rodríguez, C.W. & Vicente, C., (2005). Heterofructans and heterofructan-containing glycoproteins from sugarcane: structure and function. *Recent Research and Developmental Biochemistry*, Vol.6, pp. 31-51

Legaz, M.E.; de Armas, R., Pin˜on, D. & Vicente, C. (1998). Relationships between phenolics-conjugated polyamines and sensitivity of sugarcane to smut (*Ustilago scitaminea*). *J Exp Bot.*, Vol. 49, pp. 1723– 1728

Leu, L.S, Teng, W.S. & Wang, Z.N. (1976). Culmicolous smut of sugarcane in Taiwan, Resistant trial. Taiwan. *Sugar Exp. stn. Res. Rep.*,Vol. 74, pp. 37-45

Lloyd, H.L. & Naidoo G. (1983). Chemical array potentially suitable for determination of smut resistance of sugarcane cultivars. *Plant Dis.* Vol. 67, pp. 1103-1105

Lloyd, H.L. & Pillay,M. (1980). The development of an improved method for evaluating sugarcane resistance to smut. *Proceedings of S. Afr. Sugar Technol. Assoc. Annu. Congr.*, Vol. 54, pp. 168-172, 1980

Lovick, G. (1978). Smut of sugarcane – *Ustilago scitaminea*. *Review of Plant Pathology*, Vol. 57, No.5, pp. 181-188

Luthra , J. C.; Suttar, A. & Sandhu, S.S. (1940). Experiments on the control of smut of sugarcane. *Proceedings in Indian Academy of Sciences, Sec. B.* Vol. 12, pp. 118-128, 1940

Magarey, R.C.; Kuniata, L.S., Samson, P.R., Croft, B.J., Chandler, K.J., Irawan, Braithwaite, K.S., Allsopp, P.G., James, A.P. & Rauka, G.R. (2008). Research into exotic disease and pest threats to *Saccharum* germplasm in Australia and neighbouring countries. *Sugar Cane International*, Vol. 26, No. 1, pp. 21-25

Mameghmay, R.S. (1984). Chemotherapeutic effects of fungicides on sugarcane systemically infected by smut. *Sugarcane*, Vol. 1, pp. 3

Martinez, M.; Medina, I., Naranjo, S., Rodriguez, C.W., de Armas, R., Piñón, D., Vicente, C. & Legaz,M.E. (2000). Changes of some chemical parameters, involved in sucrose recovery from sugarcane juices, related to the susceptibility or resistance of sugarcane plants to smut (*Ustilago scitaminea*). *Int. Sugar J.*, Vol. 102, pp.445–448

Millanes, A.M.; Fontaniella, B., Legaz, M.L. & Vicente, C. (2005). Glycoproteins from sugarcane plants regulate cell polarity of *Sporisorium scitaminea* teliospores, *Journal of Plant Physiology*, Vol. 162, pp. 253-265

Millanes, A.M.; Vicente, C. & Legaz, M.E. (2008). Sugarcane glycoproteins bind to surface, specific ligands and modify cytoskeleton arragement of *Ustilago scitaminea* teliosporas. *Journal of Plant Interaction*, Vol.3, pp. 95-110

Misra, S.R.; Prasad, L., and Singh, K., (1978). Heat therapy against seed piece transmissible disease. Effect of moist hot air treatment on smut disease. *Ann. Rept. Indian Nst. of Sugarcane Research*, Lucknow

Munkacsi, A.B.; Stoxen, S. & Georgiana. (2007). Domestication of maize, sorghum, and sugarcane did not drive the divergence of their smut pathogens. *Evolution*, pp. 388-403

Nallathambi, P.; Padmanaban, P. & Mohanraj, D. (2001). Standardization of an indirect ELISA technique for detection of *Ustilago scitaminea* Syd., causal agent of sugarcane smut disease. *Journal of Mycology and Plant Pathology* Vol. 31, No.1, pp. 76-78

Nallathambi, P.; Padmanaban, P. & Mohanraj, D. (1998). Histological staining: an effective method for sugarcane smut screening. *Sugar Cane*, Vol. 2, pp. 1013

Nzioki, H.S.; Jamoza, J. E., Olweny, C. O. & Rono, J. K. (2010). Characterization of physiologic races of sugarcane smut (*Ustilago scitaminea*) in Kenya. *African Journal of Microbiology Research*. Vol. 4, No. 16, pp. 1694-1697

Olweny, C.O.; Kahiu Ngugi, Nzioki, H., Githiri, S.M. (2008). Evaluation of smut inoculation techniques in sugarcane seedlings. *Sugar Tech*, Vol. 10, No. 4, pp. 341-345

Piepenbring, M.; Stoll, M. & Oberwinkler, F. (2002). The generic position of *Ustilago maydis*, *Ustilago scitaminea*, and *Ustilago esculenta* (Ustilaginales), *Mycological Progress*, Vol.1, No. 1, pp. 71–80

Presley, J. (1978). Culmicolus smut of sugar cane: A chronology of the occurrence, spread and economic impact on the sugar cane growing countries where the disease has occurred. *Sugar y Azucar*, pp. 3439

Que, Y. X.; Lin, J.W., Zhang, J.S., Ruan, M.H., Xu, L.P., Zhang, M.Q. (2008). Molecular cloning and characterization of a non-TIR-NBS-LRR type disease resistance gene analogue from sugarcane. *Sugar Tech*, Vol. 10, pp. 71-73

Que, Y.X.; Xu, L.P., Lin, J.W., Chen, R.K. (2009). Isolation and characterization of NBS-LRR resistance gene analogs from sugarcane. *Acta Agronomica Sinica*. Vol. 35, No. 4, pp. 631–639

Raboin, L.M.; Offmann, B., Hoarau, J.Y., Notaise, J., Costet, L., Telismart, H., Roques, D., Rott, P., Glaszmann, J.C. & D'Hont, A. (2001). Undertaking genetic mapping of sugarcane smut resistance. *Proc S Afr Sug Technol Ass*, Vol. 75, pp. 94-98

Raboin, L.M.; Selvi, A., Oliveira, K.M., Paulet, F., Calatayud, C., Zapater, M.F., Brottier, P., Luzaran, R., Garsmeur, O. Carlier, J. & D'Hont, A. (2007). Evidence for the dispersal of a unique lineage from Asia to America and Africa in the sugarcane fungal pathogen *Sporisorium scitaminea*, *Fungal Genetics and Biology*, Vol. 44, pp. 64–76

Ramesh Sundar, A.; Velazhahan, R. & Vidhyasekaran, P. (2002). A glycoprotein elicitor isolated from *Colletotrichum falcatum* induces defense mechanisms in sugarcane leaves and suspension-cultured cells. *Journal of Plant diseases and Protection*, Vol.109, No.6, pp. 601-611

Riley, I. T. & Jubb, T. F. (1999). First outbreak of sugarcane smut in Australia. *Proceedings of the XXIII ISSCT Congress*, New Delhi, India 1999

Rodriguez, E.; LaO, M., Gago, S., Espino, A., Acevedo, R. & Muñiz, Y. (2001). Razas patogénicas del carbón de la caña de azúcar en Cuba. *CubaAzúcar*, Vol. 31, pp. 34–38

Santiago, R.; de Armas, R., Legaz, M.-E. & Vicente, C. (2008). Separation from *Ustilago scitaminea* of different elicitors which modify the pattern of phenolic accumulation in sugarcane leaves. *Journal of Plant Pathology*, Vol.90, No. 1, pp. 87-96

Santiago, R.; Quintana, J., Rodríguez, S., Díaz, E.M., Legaz, M.E. & Vicente C. (2010). An elicitor isolated from smut teliospores (*sporisorium scitamineum*) enhances lignin deposition on the cell wall of both Sclerenchyma and xylem in Sugarcane leaves. *Pak. J. Bot.*, 42No.4, pp. 2867-2881

Schenck, S. (2003). New race of sugarcane smut on Maui. Hawaii, Agriculture Research Center-*Pathology Report* Vol.69, pp. 1–4

Singh, A.P.; Ramji Lal & Solomon, S. (2002). Changes in Ascorbic Acid Content in Sugarcane Affected with Smut Fungus (*Ustilago scitaminea* Syd.), Vol. 4, No. l, pp. 72-73

Singh, K. & Budhraja, T. R. (1964). The role of bud scales as barriers against smut infection. *Proceedings Bien Conf Sugarcane Res Dev;* Vol. 5, pp. 687–90, 1964

Singh, N.; Somai, B.M. & Pillay, D. (2005). In vitro screening of sugarcane to evaluate smut susceptibility. *Plant Cell, Tissue and Organ Culture,* Vol. 80, pp. 259-266

Singh, N.; Somai, B.M. & Pillay, D. (2004). Smut disease assessment by PCR and microscopy in inoculated tissue cultured sugarcane cultivars. *Plant Science,* Vol. 167, pp. 987–994

Sinha, O.K. & Singh, K. (1982). Stain technique for detection of smut hyphae in buds of sugarcane. *Plant Disease,* Vol.66, No. 10, pp. 932-933

Srinivasan, K.V. and Rao, J.T. (1968). Hot water treatment of sugarcane seed material - a simple method for rural areas. Indian Farming, Vol. 18, pp. 25

Sydow, H. (1924). Notizen Uber Ushlagineen. *Ann. Mycol.,* Vol.22, pp. 277

Thokoane, L.N. & Rutherford, R.S. (2001). CDNA-AFLP Differential display of sugarcane (*Saccharum* Spp. Hybrids) genes induced by challenge with the fungal pathogen *Ustilago scitaminea* (Sugarcane Smut). *Proceedings of S Afr Sug Technol Ass.,* Vol. 75, pp. 104-107, 2001

Wada, A. C. (2003). Control of sugarcane smut disease in Nigeria with fungicides. *Crop Protection,* Vol. 22, No.1, pp. 4549

Wada, A.C.; Mian, M.A.W., Anaso, A.B., Busari, L.D. & Kwon-Ndung, E.H. (1999). Control of Sugarcane Smut (*Ustilago scitaminea* Syd) Disease in Nigeria and Suggestions for an Integrated Pest Management Approach, Control of Sugarcane Smut (*Ustilago scitaminea* Syd) Disease in Nigeria and Suggestions for an Integrated Pest Management Approach. *Sugar Tech.,* Vol.1, No. 3, pp. 48 - 53

Waller, J.M. (1970). Sugarcane smut (*Sporisorium scitaminea*) in Kenya. II. Infection and resistance. Trans British Mycol Soc; 54:405–14.

Wang, J.N.; Xu, L.P. & Chert, R.K. (1995). A preliminary study on pathological mechanisms of sugarcane chlorotic streak. *Sugarcane,* Vol. 2, No.4, pp. 10-13

Xu Liping; Wang Jiannan & Chen Rukai. (1994). Biochemical Reaction of Sugarcane to Smut and Its Relation to Resistance. *Sugarcane* Vol. 3

Xu, L.; Que, Y. & Chen, R. (2004). Genetic Diversity of *Ustilago scitaminea* in Mainland China. *Sugar Tech.* Vol. 6, No. 4, pp. 267 – 271

Yudilai Muñiz, B.; Martínez & María, La O. (2004). Sugarcane smut (*Ustilago scitaminea* Sydow): diagnostic methods. *Rev. Protection Veg.,* Vol. 19, No. 1, pp.

5

Physico-Chemical, Biochemical and Microbiological Phenomena of the Medicinal and Aromatic Plants Extract Used in the Preparation of *Tassabount* Date Juice in Morocco

Hasnaâ Harrak[1], Marc Lebrun[2],
Moulay Mustapha Ismaïli Alaoui[3],
Samira Sarter[2] and Allal Hamouda[3]
[1]National Institute of Agricultural Research, Marrakesh
[2]International Cooperation Centre of
Agricultural Research for Development, Montpellier
[3]Hassan II Agricultural and Veterinary Institute, Rabat
[1,3]Morocco
[2]France

1. Introduction

In the Moroccan oases, traditional preparations of dates, the fruits of date palm (*Phoenix dactylifera*), are often associated with medicinal and aromatic plants (MAPs) which provide the properties of flavoring, preservation and medication (Harrak, 2007). For the traditional dates juice, *Tassabount*, its nutritional and organoleptic qualities and its therapeutic virtues come from both the date genotypes (cultivars and wild hybrids) and a multitude of MAPs used in its preparation (Harrak et al., 2009). Considering its promising applications, the *Tassabount* juice can get out of household manufacturing and consumption to emerge as a local product for a wider market. Such valorization requires a deep description and understanding of the different steps of the traditional juice processing.

Harrak et al. (2009) described the household process of preparing *Tassabount* which consists of two main steps. The first one is to prepare an aqueous extract of MAPs. This step is commonly called "fermentation" by households. About thirty MAPs can be used for preparing *Tassabount* (Zirari et al., 2003; Harrak, 2007). These are crop or wild plants used to make beverages, perfumes, medicinal extracts and herbal teas and tea bags (Table 1). Made empirically, these MAPs confer various therapeutic virtues and aromatic notes to *Tassabount* juice. The second step is juice processing. A viscous mixture is made with dates and a gradual incorporation of the MAPs extract. The rich foam reminiscent of soap forming, gave the name *Tassabount*. The mixture is finely sieved to remove seeds and part of the pulp retentate (Harrak et al., 2009).

Plants	Scientific names[1]
Basil	*Ocimum basilicum* L.
Bitter almond	*Prunus amygdalus* Stokes
Buttercup	*Pulicaria arabica* (L.) Cass.
Clove	*Eugenia cariophyllata* Thunb. (*Syzygium aromaticum* (L.) Merr.)
Date palm (fruits: dates)	*Phœnix dactylifera* L.
Fumitory	*Fumaria capreolata* L. / *Fumaria officinalis* L. / *Fumaria agraria* Lag. / *Fumaria parviflora* Lam. / *Euphorbia obtusifolia* Poiret / *Euphorbia helioscopa* L.
Gaillonia	*Gaillonia reboudiana* Coss. et Dur.
Haloxylon	*Haloxylon scoparium* Pomel
Harmel (roots)	*Peganum harmala* L.
Henna (leaves)	*Lawsonia inermis* L. (*Lawsonia alba* Lamk.)
Hundred petaled rose	*Rosa damascena* Mill. / *Rosa centifolia* Mill.
Iris (roots)	*Iris germanica* L. / *Iris florentina* L.
Lemon (fruits: lemons)	*Citrus limon* (L.) Burm.
Lime (fruits: limes)	*Citrus limon* (L.) Burm. / *Citrus aurantiifolia* Swingle
Mandrake	*Mandragora autumnalis* L.
Mint round	*Mentha suaveolens* Ehr. (*Mentha Rotundifolia* (L.) Hudson)
Myrtle	*Myrtus communis* L.
Nutmeg	*Myristica fragrans* Houtt
Oregano	*Origanum compactum* Benth. / *Origanum vulgare* L.
Ormenis	*Ormenis africana* Jord. et Fourr. / *Ormenis scariosa* (Ball.) Lit. et Maire
Pennyroyal	*Mentha pulegium* L.
Rosemary	*Rosmarinus officinalis* L.
Round shoveler	*Cyperus rotondus* L.
Sagebrush	*Artemisia herba-alba* Asso.
Sarghine (roots)	*Corrigiola telephiifolia* Pour.
Thyme	*Thymus satureioides* Coss. et Ball. / *Thymus broussonetii* Boiss. / *Thymus pallidus* Coss. / *Thymus maroccanus* Ball. / *Thymus vulgaris*
Zygophylle (roots)	*Gactulum album* / *Zygophyllum gaetulum* Emb. et Maire *Zygophyllum waterloti* Maire / *Zygophyllum fontanesi* Webb.

[1] (Sijelmassi, 1996).

Table 1. Medicinal and aromatic plants used in the preparation of *Tassabount* dates juice.

Depending on physical, chemical and microbiological parameters of MAPs extract, this step is critical to the *Tassabount* quality. The aim of this work is to provide a better understanding of the main physical, physico-chemical, biochemical and microbiological phenomena that take place during the preparation of MAPs aqueous extract. Nutritional, organoleptic and hygienic qualities of this extract are evaluated as well as its impact on the quality of *Tassabount* for a better valorization of this juice.

2. Materials and methods

2.1 Standardization of the aromatic extract

To set the conditions for preparation of standardized aqueous extract, preliminary tests for the preparation of aromatic extracts were performed. These trials included the selection of MAPs, temperature and maceration time.

2.1.1 Selection of medicinal and aromatic plants

Six plants were chosen among the most used by the oases households in the preparation of *Tassabount*. They are known for their antiseptic, antispasmodic, diuretic, carminative, antibacterial and / or antioxidant proprieties. These are oregano (*Origanum vulgare* L.), thyme (*Thymus vulgaris* L.), hundred petaled rose (*Rosa centifolia*), pennyroyal (*Mentha pulegium* L.), sarghine (*Corrigiola telephiifolia*) and harmel (*Peganum harmala* L.) (Fig. 1).

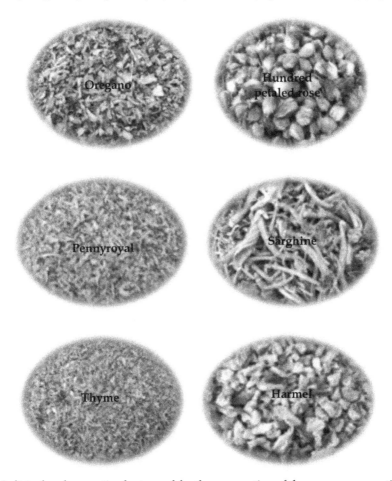

Fig. 1. Medicinal and aromatic plants used for the preparation of the aqueous aromatic extract.

2.1.2 Medicinal and aromatic plants concentration

In general, concentrations of MAPs used in traditional food or pharmaceutical preparations are based on the know-how of the oases population who experienced from the earliest times how to control the active ingredients and to avoid harmful overdoses.

Based on the experience of oases women, the quantities of plants used for a volume of five liters of water were set as follows: 25 g of origano, 25 g of hundred petaled rose, 25 g of thyme, 25 g of pennyroyal, 12.5 g of sarghine roots and 12.5 g of harmel roots (Harrak, 2007). The amount used of harmel roots, much less rich in alkaloids than seeds, is tiny compared to the doses known to induce toxicity (Hammiche & Merad, 1997, as cited in Ben Salah et al., 1986). However, awareness and information on the toxic potential of plants, used in various traditional preparations, are an important preventive measure.

Before use, the MAPs are first sifted, sorted, washed and drained. Three dates of the *Black Bousthammi* variety (Fig. 2) chosen for the preparation of the juice were added as a source of carbohydrates and aromas in the MAPs maceration step.

Fig. 2. Dates of *Black Bousthammi* variety during ripening (on the left) and mature (on the right).

2.1.3 Temperature and duration of medicinal and aromatic plants maceration

Oases women immerse plants in water at room temperature and let marinate for two to three days. This period is deemed sufficient by the women for both an effective diffusion of plants constituents in the aqueous phase and for a good "fermentation". However, some women, yielding to the facility and depleting PAMs, use extracts leftover exceeding 5 days of maceration, accepting therefore a decline in the quality. To this end, we felt more appropriate in our work to extend the maceration beyond three days to detect signs of possible alteration of the quality of the extract.

2.2 Characterization of physical, physico-chemical and biochemical phenomena of medicinal and aromatic plants extract

2.2.1 Preparation of the medicinal and aromatic plants extract

Three jars, each containing water and aliquots of the six plants as described above, were covered with a cloth (not airtight closure) and placed at an average ambient temperature of 23 °C +/- 1 °C for monitoring the maceration.

Physico-Chemical, Biochemical and Microbiological Phenomena of the Medicinal and
Aromatic Plants Extract Used in the Preparation of Tassabount Date Juice in Morocco

111

The different physico-chemical and biochemical analyses were performed on filtered extracts collected from the three jars during five days or more.

2.2.2 Physical and physico-chemical criteria

1. Weight loss: The weight loss of the extract, providing information on gas release and / or water evaporation during the MAPs maceration, was followed by the weighing of the three jars. It is expressed in %.
2. Brix: The Brix of the extract during the MAPs maceration was determined at 20 °C using a digital refractometer (Pocket Refractometer PAL-1 (0~53%), ATAGO). It is expressed in °Bx (AOAC, 1990).
3. pH: The pH of the extract during the MAPs maceration was measured at 20 °C using a pH meter (SCHOTT) on a sample of 20 ml of filtered extract, under continuous stirring (AOAC, 1990).
4. Total titratable acidity: The total titratable acidity of the extract during the MAPs maceration was determined by titration of 20 ml of filtered extract using TitroLine Easy (SCHOTT) to pH 8.1 with a solution of sodium hydroxide 0.1 N. The acidity is expressed in meq / 100 ml (AOAC, 1990).
5. Ultra violet / visible spectrum: The evolution of the absorbance of the MAPs extract according to the wavelength from 190 nm to 900 nm (UV / Visible zone) was determined at four maceration times (20 min, 78 h, 102 h and 212 h). The measurements were made with a UV / visible spectrophotometer (UVIKON 933 Spectrophotometer Double Beam UV/VIS) using quartz cells with optical path equal to one centimeter. The color intensity of the extract during the MAPs maceration was determined as the sum of the luminous absorbance at 420, 520 and 620 nm.

2.2.3 Biochemical criteria

1. Total polyphenols: The total polyphenols contained in the aqueous extract during the MAPs maceration were extracted by an acetone / water mixture. They were determined by the Folin-Ciocalteau method (colorimetric method) revealing a blue color. Elimination of molecules with reducing properties, disturbing the determination of polyphenols by the Folin-Ciocalteau method, was performed on a cartridge with absorptive capacity of the polyphenols. Elution of the polyphenols was carried out with methanol. Quantification was performed using an external calibration of gallic acid at 760 nm and was expressed in mg GAE (Gallic Acid Equivalent) per 100 g of aqueous extract (Georgé et al., 2005). Measurements of absorbance were performed using a UV / Visible spectrophotometer (UVIKON 933 Spectrophotometer).
2. Ethanol: The dosage of ethanol of MAPs extract was achieved using an Enzytec™ fluid ethanol Kit for the photometric determination of ethanol. The samples were introduced into small cells of PS type for spectrophotometer. Measurements of absorbance at the wavelength of 340 nm were performed using a UV / Visible spectrophotometer (UVIKON 933 Spectrophotometer). The ethanol content was expressed in g / l of the aqueous extract of plants.
3. Lactic acid: The dosage of lactic acid of MAPs extract was achieved by using an Enzytec™ L-lactic Acid/D-lactic acid Kit for the photometric determination of lactic acid. The samples were introduced into small cells of PS type for spectrophotometer.

Measurements of absorbance at the wavelength of 340 nm were performed using a UV/ Visible spectrophotometer (UVIKON 933 Spectrophotometer). The lactic acid content was expressed in g / l of aqueous extract of plants.

2.2.4 Production of carbon dioxide

The release of carbon dioxide during the maceration of plants was followed in three sealing jars. The concentrations of CO_2 and O_2 in the jars, expressed in %, were measured using an O_2/CO_2 analyzer (PBI Dansesor, Checkmate 9900).

2.2.5 Evolution of the aromatic profile

The aromatic profile of the extract was followed during the MAPs maceration by implementing an experimental system based on Solid Phase Micro Extraction (SPME, SUPELCO Inc). This technique allows direct sampling and concentration, of the volatile emissions present in the headspace above the aqueous extract. The aromatic profile changes were monitored during three days by taking samples every two hours.

Trapping of the volatile aromatic fraction of the extract was performed using a manual fiber holder for 20 minutes. Volatile substances adsorbed by the fiber of polydimethylsiloxane type (PDMS) were then injected and desorbed thermally at 250 °C.

Separation and identification of volatile compounds were carried out by gas chromatography coupled with mass spectrometry (Agilent 5973N) using (J&W, DB-WAX) capillary column 0.25 mm (inner diameter) x 30 m x 0.25 μm (film thickness). The flow rate of carrier gas (helium) was 1.0 ml / min. Oven temperature programmation was from 40 °C to 220 °C for 10 min, at the rate of 3 °C / min (Fig. 3).

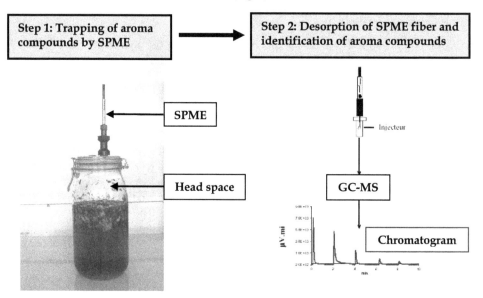

Fig. 3. Experimental system monitoring the aromatic potential of medicinal and aromatic plants extract.

Kovats retention indices were generated by using a series of n-alkanes from C-5 to C-22 eluted under the same chromatographic conditions as samples of aromatic extract.

The acquisition, visualization and analysis of data were made using the MSD ChemStation software (Agilent) and the NIST 2.0 a library of reference mass spectra.

2.2.6 Global olfactory fingerprint

The study of the aromatic profile during the maceration was also performed on the aqueous extract of MAPs by monitoring the global olfactory fingerprint using the electronic nose. A total of 15 samplings were carried out during the five days of maceration. At each sampling, 6 replicates of 3 ml of extract were withdrawn using a precision syringe and introduced into 10 ml chromacol headspace vial then immediately crimped.

The headspace was generated by incubation at 60 °C during 15 min with stirring (500 rpm). 2 ml of headspace were sampled and injected at a rate of 1.5 ml/s. They were injected (flow rate of purified air of 150 ml / min) in the Prometheus electronic nose (Alpha MOS) equipped with 18 MOS (metal oxide sensor). The sensor response, based on adsorption-desorption phenomenon, depends on the affinity and concentration of the molecules present in the head space. Time acquisition was 2 minutes, followed by 8 minutes of relaxation for a correct return to baseline.

So, each sample was associated to 18 different values of the sensors, considered as variables. They were analyzed by factorial discriminant analysis (FDA) using the Prometheus Alphasoft software (version 7), generating a global olfactory fingerprint which is considered as a real identity card.

2.3 Microbiological evolution of the extract

2.3.1 Aromatic extract sampling

A jar containing water and plants, prepared by the same method as the standard extract, was sealed to prevent outside contamination. An amount of 9 ml of extract was taken from the aqueous portion under stirring at four different times of plants maceration (0 h, 24 h, 48 h and 72 h) and then it was filtered. The removal and filtration of the extract were carried out in aseptic conditions (in laminary flow microbiological cabinet).

2.3.2 Enumeration of the microbial flora

Culture media: PCA (Plate Count Agar, Bio Merieux), PDA (Potato Dextrose Agar, Bio Merieux) and MRS (Man Agar, Rogosa, Sharpe) were respectively used to monitor the evolution of total mesophilic microflora, fungal microflora and lactic microflora. Serial dilutions up to 10^{-6} of the extract were performed using sterile physiological water (containing 0.85 % NaCl). For PCA and PDA media, 1 ml of extract was inoculated in the mass and incubated at 30 °C for 48 h. For the MRS medium, 0.1 ml was spread on the surface using sterile glass balls. The Petri dishes were then placed in airtight jars containing a Genbox anaer bag (generator for the cultivation in the jar of anaerobic bacteria) (Bio Merieux). Monitoring of anaerobic conditions was achieved using a moistened indicator paper. Incubation was carried out at 37 °C during 72 h. Control dishes (extract free) were also prepared for the three culture media. The results were expressed as logarithm (Log) of colony-forming units (cfu) per liter of aqueous extract of plants.

3. Results and discussion

3.1 Evolution of physical and physico-chemical criteria of aromatic extract

3.1.1 Weight loss

Fig. 4 shows the evolution of weight losses of MAPs extract during the maceration, calculated from the initial weight. Weight losses during the first four days of maceration were respectively 0.59 %, 1.12 %, 1.52 % and 2.03 %, reaching 4.08 % after eight days of maceration. This weight loss could be mainly explained by gas release during the maceration.

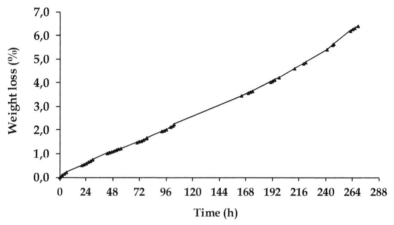

Fig. 4. Weight loss (%) of the aromatic extract during the maceration of medicinal and aromatic plants.

3.1.2 Brix

The largest change of Brix took place during the first twenty-eight hours of maceration (Fig. 5). In fact, the Brix value doubled during this time, from 0.4 °Bx to about 0.9 °Bx and this demonstrated the migration, into the aqueous phase, of soluble solids including sugars present in plants and the dates pulp added. This was followed by stabilization at this value during the second day, then a slight decrease during the third day. This decrease could be due to the use of sugars by microorganisms for their metabolism.

3.1.3 pH and total titratable acidity

The pH was about 7.3 at the beginning of the maceration and decreased steadily to the value of 5.3 after 44 h (Fig. 6). Then, the pH tended to stabilize at this value. The acidity followed the opposite trend. It recorded at the beginning a value of 0.8 meq / 100 ml, which increased with maceration time to reach 5.8 meq / 100 ml after 44 h. This increase can be explained by the production of organic acids during this MAPs maceration step. After this time, measures stabilized.

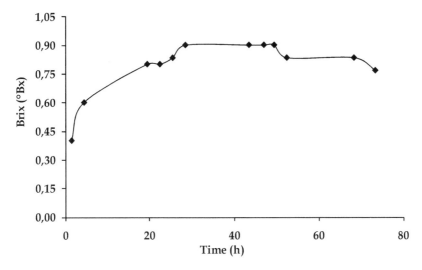

Fig. 5. Evolution of the aromatic extract Brix during the maceration.

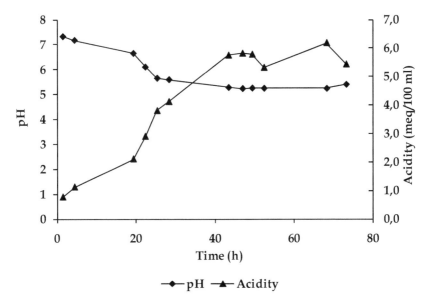

Fig. 6. Evolution of pH and total titratable acidity of the aromatic extract during the maceration.

The aromatic extract was slightly acidic (pH near 5), which could inhibit the growth of certain pathogenic bacteria. In practice, some households use, among MAPs, pieces of lemon or lime to increase the acidity of the extract.

3.1.4 UV / Visible spectrum

The evolution of extract UV / Visible spectrum during the maceration of MAPs showed the presence of peaks at some wavelengths. The presence of these peaks indicated release into water of some compounds of MAPs as phenolic compounds. Three major peaks appeared in the UV region at wavelengths: 338 nm, 281 nm and 197 nm (main peak). The latter was also observed in the extract just at the beginning of the plants maceration (after 20 min). The two spectra taken after 3 days (102 h and 212 h) coincided perfectly, indicating about a saturation of the liquid extract and / or depletion of MAPs. These two spectra were close and had the same pace as that recorded after a maceration time of 78 h. Only the peak intensities differed, heralding a lower concentration in different components of the extract at 78 h compared to those recorded after three days. The spectrum taken at the beginning of maceration had the same profile as those taken at the end of maceration, while showing very significant differences in intensity, so in concentration of absorbing compounds at these different wavelengths. This indicated a rapid diffusion of these compounds in water (Fig. 7).

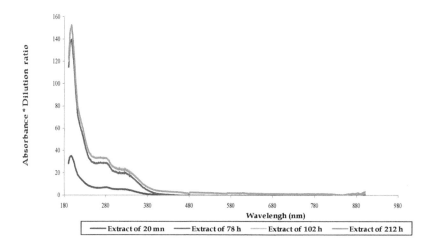

Fig. 7. Spectra of the extract in UV/Visible obtained at different times of maceration of aromatic and medicinal plants.

Moreover, a similar study on the kinetics of the absorbance at different wavelengths of green tea extracts also revealed an increase in absorbance at 280, 330 and 350 nm with time of infusion in water (Nkhili et al., 2007). The compounds corresponding to these peaks were, respectively, total phenols, hydroxycinnamic derivatives and flavonols.

Concerning the color intensity, it increased slightly during the first three days of maceration (from 0.164 recorded at 20 min to 0.797 recorded after 3 days), and recorded a strong increase (up to 7.200) in the fourth day to be stable thereafter. This increase could be due to the diffusion of colorants and to the browning of the extract due to oxidative phenomena. Its stabilization could be attributed to a depletion of plants in these compounds and eventually to an equilibrium established between the aqueous phase and plants. This coloration of the extract, however, did not affect the brown color of the dates juice.

3.2 Evolution of biochemical criteria of aromatic extract

3.2.1 Total polyphenols

The total polyphenols content increased during the first three days of maceration of plants and tended to stabilize at a value of about 100 mg gallic acid equivalent (GAE) / 100 g of extract after the fourth day. These values were in fact 24.21, 95.00 and 99.98 mg GAE / 100 g extract, respectively, after 20 min, 78 h and 102 h of plants maceration.

3.2.2 Ethanol

The evolution of the ethanol content was fast at the beginning of maceration: the concentration had indeed increased from zero, recorded at 20 min, to 2.17 g / l, recorded after the third day. Between the third and the eighth days of maceration, we recorded a slight increase, reaching only 0.7 g / l. However, we may retain that the content of the aromatic extract ethanol was generally low, even after eight days of maceration.

3.2.3 Lactic acid

The lactic acid concentration increased with the duration of maceration. It reached 0.92 g / l after three days (78 h) and 1.30 g / l after almost nine days (212 h). The presence of lactic acid in the extract is an asset for the stabilization of the *Tassabount* juice. Indeed, lactic acid is widely used as a preservative in foods. Its combination with essential oils of MAPs could be valuable to preserve the sensory quality of juice and to prevent pathogens contamination that might require a higher acidity for their destruction or their limitation of growth (Dimitrijević et al., 2007).

3.3 Evolution of carbon dioxide

The evolution of carbon dioxide production during the maceration of plants for a period of ten days is shown in Fig. 8. It is clear that the production of carbon dioxide increased with time reaching a maximum average of 76.6 % after one week. The anaerobiosis was reached after 46 h. This change in the reverse direction of carbon dioxide (CO_2) and oxygen (O_2) indicated a fermentative process occurring during maceration of MAPs. The intense release of CO_2 during the first four days was consistent with the weight loss of the aromatic extract recorded during the maceration of MAPs.

It should be reminded that the monitoring of concentrations of both gases was performed in airtight conditions. The gas phase above the extract, estimated to be about one third of the total volume, is very important for the interparticle transfer of oxygen and carbon dioxide generated by the metabolism thus affecting the microbial metabolism (Desgranges & Durand, 1990).

Furthermore, the increase of CO_2 concentration during the maceration of MAPs generated a pressure in the gas phase. Previous studies have shown that the partial pressure of CO_2 had an effect on the physiological behavior of filamentous fungi by inhibiting their growth (Desgranges & Durand, 1990).

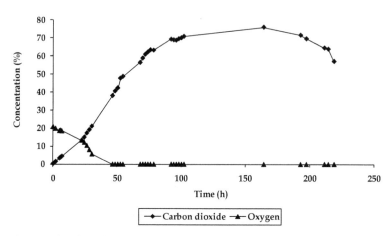

Fig. 8. Evolution of carbon dioxide and oxygen concentrations (%) during the maceration of medicinal and aromatic plants.

3.4 Evolution of the aromatic profile of plants extract

3.4.1 Quantitative and qualitative analysis of aroma compounds

A full qualitative analysis, taking into account all of the volatile compounds extracted by SPME, has identified a total of 92 volatile compounds in the aqueous extract of MAPs (Table 2). Of these, 96.7 % were identified in the extract taken after 6 h of maceration of plants, against only 54.3 % for the extract taken after 48 h. The compounds known for their functional properties such as thymol, carvacrol and α-pinene, were identified in both situations. These qualitative observations suggested that the maceration time of 6 h allowed the best expression of aromatic plants.

The 92 identified aroma compounds showed the predominance of terpenic hydrocarbons and their alcoholic and ketonic derivatives. Terpenic esters, oxygenated sesquiterpenes, aliphatic alcohols, aliphatic ketones and aldehydes were found as well.

For some aromatic compounds identified by SPME, the interpretation of results was based on a semi-quantitative approach, by determining the dominant compounds (with relatively large peaks), compounds with low peaks and compounds that were not detected in the extract at different studied maceration times (Table 3). The data in this table show mainly quantitative variations of the compounds at different times of maceration. The qualitative composition of the extract at these different times remains substantially the same with a few exceptions, where we record the absence or presence at a low concentration. The majority of compounds having low Kovats indices (ranging from 997 to 1243) were identified in the headspace, in important amounts, at the beginning of maceration (recorded after 20 min, the requisite trapping time) until the third day. Other compounds were also found at the beginning of maceration, this was the case especially of β-linalool, (-)-bornyl acetate, thymyl methyl ether, β-caryophyllene, isoborneol, thymol and carvacrol.

However, some compounds had a random detection during the maceration as cis-geranylacetone and mint furanone. This could be due to saturation of the SPME fiber or possibly they have competed with the other compounds. As for the humulene oxide, it has been appeared after the sixth hour of maceration.

N°	Aromatic compound	Kovats index	Extract of 6 h[2]	Extract of 48 h[2]
1	Tricyclene	997	+	+
2	Alpha-thujene	1010	+	+
3	Alpha-pinene	1012	+	+
4	Camphene	1045	+	+
5	Cyclobutanol[1]	1060	-	+
6	Beta-pinene	1088	+	+
7	Alpha-phellandrene[1]	1104	+	+
8	3-heptanone[1]	1126	+	+
9	Delta-3-carene[1]	1129	+	+
10	Beta-myrcene	1143	+	+
11	Pseudolimonene[1]	1148	+	+
12	Alpha-terpinene	1158	+	+
13	D-limonene	1178	+	+
14	Beta-phellandrene	1186	-	+
15	Eucalyptol	1190	+	+
16	Beta-*trans*-ocimene	1215	+	-
17	Gamma-terpinene	1225	+	+
18	3-octanone	1228	+	+
19	Beta-*cis*-ocimene	1230	+	+
20	*p*-cymene	1243	+	+
21	Terpinolene	1259	+	+
22	Delta-4-carene[1]	1264	+	+
23	3-methylcyclohexanone[1]	1286	+	-
24	2-octanol[1]	1295	+	-
25	3-octylacetate[1]	1315	+	+
26	*Cis*-Rose oxide	1328	+	-
27	3-nonanone	1331	+	-
28	*Trans*-Rose oxide	1341	+	-
29	L-fenchone[1]	1361	+	-
30	Nonanal	1366	+	-

N°	Aromatic compound	Kovats index	Extract of 6 h[2]	Extract of 48 h[2]
31	3-octanol	1369	+	+
32	Thujone[1]	1386	+	-
33	Beta-thujone[1]	1404	+	-
34	Linalool oxide[1]	1410	+	-
35	1-octen-3-ol[1]	1420	-	+
36	*Trans*-limonene oxide[1]	1423	+	-
37	Isomenthone	1429	+	+
38	*Trans*-sabinenehydrate	1436	+	+
39	Alpha-cubebene[1]	1442	+	-
40	Menthofurane[1]	1446	+	+
41	Alpha-campholenal[1]	1451	+	-
42	D-isomenthone	1455	+	+
43	Camphore	1471	+	+
44	Alpha-copaene[1]	1475	+	-
45	Alpha-bourbonene[1]	1499	+	-
46	Alpha-gurjunene[1]	1510	+	-
47	Beta-linalool	1517	+	+
48	Pinovarvone[1]	1521	+	-
49	Bergamol[1]	1527	+	-
50	*Trans*-isopulegone	1529	+	+
51	Bornyl formate	1535	+	+
52	Isopulegone	1542	+	+
53	(-)-Bornyl acetate	1548	+	+
54	Menthol acetate[1]	1550	+	-
55	Thymol methyl ether[1]	1558	+	-
56	Dihydrocarvone	1565	+	-
57	Thymyl methyl ether	1567	+	+
58	Beta-caryophyllene	1571	+	+
59	Thujanol	1575	+	+
60	Carvone[1]	1582	+	-
61	Delta-elemene[1]	1585	+	-
62	(+)-isomenthol[1]	1597	+	-
63	Pulegone	1601	+	+
64	*Trans*-pinocarveol(1)	1606	+	-

N°	Aromatic compound	Kovats index	Extract of 6 h[2]	Extract of 48 h[2]
65	Aromadendrene(1)	1610	+	-
66	Estragole(1)	1611	+	-
67	Cinerone	1614	+	+
68	Cis-verbenol[1]	1619	+	+
69	Alpha-caryophyllene[1]	1620	+	-
70	Isoborneol	1655	+	+
71	Alpha-terpineol	1662	+	+
72	D-carvone	1666	+	-
73	Neral	1668	+	-
74	Beta-bisabolene	1688	+	-
75	(R)-(+)-beta-citronellol	1718	+	+
76	Alpha-curcumene[1]	1728	+	-
77	Anethole	1757	+	+
78	Cis-carveol[1]	1770	+	-
79	p-cymen-8-ol[1]	1775	+	-
80	Nerol	1785	+	-
81	Cis-geranylacetone[1]	1795	+	-
82	Beta-phenylethanol	1821	+	+
83	Piperitone	1833	+	+
84	Caryophyllene oxide	1900	+	+
85	Cinerolone[1]	1903	+	-
86	Methyleugenol[1]	1923	+	-
87	Humulene oxide	1960	+	+
88	Spathulenol[1]	1975	+	-
89	Eugenol[1]	1994	+	-
90	Thymol	2045	+	+
91	Carvacrol	2064	+	+
92	Mint furanone[1]	2213	+	-
Total of identified compounds			89	50

[1] Identification attempt based on Kovats retention indices on polar column, the spectra reference of NIST library and the comparison with mint chromatograms.

[2] +: Presence of the aromatic compound; -: absence of the aromatic compound.

Table 2. Volatile compounds identified in the extract of medicinal and aromatic plants after 6 h and 48 h of maceration.

Aromatic compound	Kovats index	Maceration time (h)[1]															
		0	2	4	6	8	23	25	27	29	31	48	50	52	54	56	71
Tricyclene	997	*	*	*	+	*	+	*	+	+	+	+	+	+	+	+	+
α-thujene	1010	+	+	+	+	+	+	+	+	+	+	+	+	+	+	+	+
α-pinene	1012	*	*	*	+	+	+	+	+	+	+	+	+	+	+	+	+
Camphene	1045	+	+	+	+	+	+	+	+	+	+	+	+	+	+	+	+
β-pinene	1088	*	+	*	+	+	+	+	+	+	+	+	+	+	+	+	+
β-myrcene	1143	*	+	*	+	+	+	+	+	+	+	+	+	+	+	+	+
α-terpinene	1158	*	+	*	+	+	+	+	+	+	+	+	+	+	+	+	+
D-limonene	1178	*	+	*	+	+	+	+	+	+	+	+	+	+	+	+	+
β-phellandrene	1186	*	*	*	-	-	*	*	*	+	*	*	+	+	*	+	*
Eucalyptol	1190	*	*	*	+	+	+	*	*	+	*	*	*	+	*	*	*
γ-terpinene	1225	+	+	+	+	+	+	+	+	+	+	+	+	+	+	+	+
3-octanone	1228	*	*	*	+	*	*	*	*	+	*	*	+	+	+	+	+
p-cymene	1243	+	+	+	+	+	+	+	+	+	+	+	+	+	+	+	+
3-octanol	1369	*	*	*	+	*	*	*	*	*	*	*	*	*	*	*	*
Isomenthone	1429	*	*	*	+	*	+	*	*	+	*	*	*	*	*	*	*
Trans-sabinenehydrate	1436	*	*	*	+	*	*	*	*	*	*	*	*	*	*	*	*
D-isomenthone	1455	*	*	*	+	*	*	*	*	*	*	*	*	*	-	*	*
Camphor	1471	*	*	*	+	*	*	*	*	*	*	*	*	*	-	*	*
β-linalool	1517	+	+	*	+	+	+	+	+	+	+	+	+	+	+	+	+
Trans-isopulegone	1529	*	*	*	+	*	*	*	*	*	*	*	*	*	-	*	*
Bornyl formate	1535	*	*	*	+	*	+	*	*	*	*	*	*	*	-	*	*
Isopulegone	1542	*	+	*	+	+	+	+	+	+	*	*	*	+	-	+	*
(-)-Bornyl acetate	1548	+	+	*	+	+	+	+	+	+	+	+	+	+	+	+	+
Dihydrocarvone	1565	*	*	*	+	*	*	*	*	*	*	*	*	*	*	*	*
Thymyl methyl ether	1567	+	+	*	+	+	+	+	+	+	+	+	+	+	+	+	+
β-caryophyllene	1571	+	+	*	+	+	+	+	+	+	*	*	+	+	*	+	*
Thujanol	1575	*	*	-	+	*	*	*	*	*	*	*	+	+	*	+	*
Pulegone	1601	+	+	+	+	+	+	+	+	+	+	+	+	+	+	+	+
Cinerone	1614	*	*	-	+	*	*	*	*	*	*	*	*	*	*	*	*
Isoborneol	1655	+	+	+	+	+	+	+	+	+	+	+	+	+	+	+	+
α-terpineol	1662	*	*	*	+	*	*	*	*	*	*	*	*	*	*	*	*
β-bisabolene	1688	*	*	*	+	-	-	*	*	*	*	*	*	*	*	*	-
(R)-(+)-β-citronellol	1718	*	*	*	+	*	*	*	*	*	*	*	*	*	*	*	*

Aromatic compound	Kovats index	Maceration time (h)[1]															
		0	2	4	6	8	23	25	27	29	31	48	50	52	54	56	71
Anethole	1757	*	*	*	+	*	*	*	*	*	*	*	*	*	*	*	*
Cis-geranylacetone	1795	-	*	-	+	*	*	-	-	*	-	-	*	*	*	*	-
β-phenylethanol	1821	*	*	*	+	*	*	*	*	*	*	*	*	*	*	*	*
Piperitone	1833	*	*	*	+	*	*	*	*	*	*	*	*	*	*	*	*
Caryophyllene oxide	1900	*	*	-	+	*	*	*	*	*	*	*	*	*	*	*	*
Humulene oxide	1960	-	-	-	+	*	*	*	*	*	*	*	*	*	*	*	*
Thymol	2045	+	+	*	+	+	+	+	+	+	+	+	+	+	+	+	+
Carvacrol	2064	+	+	+	+	+	+	+	+	+	+	+	+	+	+	+	+
Mint furanone	2213	-	*	-	+	*	*	-	*	*	*	-	*	*	-	*	-

[1] + Major presence; * Weak presence; - "Analytical" absence.

Table 3. Main aroma compounds identified in the aromatic extract during the maceration of the medicinal and aromatic plants.

3.4.2 Functional properties of the main aroma compounds

Carvacrol, with an important flavoring power, is approved by Food and Drug Administration (FDA - USA) for food use. It was included by the European Union in the list of category B chemical aroma that can be added to food products, for example 2 ppm in drinks and 25 ppm in candy (De Vincenzi et al., 2004). Several studies have shown that carvacrol has several biological activities. It is anthelmintic, antibacterial, antidiuretic, anti-inflammatory, antioxidant, antiseptic, antispasmodic, antitussive, expectorant, carminative, fungicidal, irritating, pesticide and worming (Akrout, 2004, as cited in Duke, 1998). Just like carvacrol, thymol has strong antioxidant activities comparable to those of known antioxidants such as α-tocopherol and butylated hydroxytoluene (BHT). Therefore, the ingestion of these aroma compounds can help prevent *in vivo* oxidation damages such as lipid peroxidation that is associated with cancer, premature aging, atherosclerosis and diabetes (Lee et al., 2005).

The α-pinene, present among the major compounds of the PAMs extract and among the aroma compounds of Moroccan dates as well (Harrak et al., 2005), has several biological activities: it is antibacterial, antiviral, anti-inflammatory, expectorant, sedative, herbicide, insect repellent and flavoring (Akrout, 2004, as cited in Duke, 1998).

In addition, numerous investigations have confirmed the antimicrobial action of essential oils in model food systems and in real foods. In this regard, many studies have shown that the essential oils of oregano (*Origanum vulgare*) and thyme (*Thymus vulgaris*) are among the most active ones against a number of food spoilage and pathogen microorganisms (Dimitrijević et al., 2007).

3.5 Global olfactory fingerprint evolution changes

Factorial discriminant analysis was applied to the rectangular table of data whose rows are samples of aromatic extracts divided into groups by time of maceration and whose columns are the responses of 18 sensors of the electronic nose. For each of the available 15 periods of maceration (groups), the six extract samples gave us a table of dimension 90 x 18. Maceration times studied were: 0 h, 2 h, 5 h, 18 h, 24 h, 26 h, 29 h, 42 h, 46 h, 49 h, 52 h, 69 h, 73 h, 76 h and 139 h. The goal was to find the discriminant axes separating the best possible in projection of the 15 groups based on the responses of the electronic nose sensors. In this analysis, the first three axes explained respectively 91.4 %, 4.4 % and 3.8 % of the total variance.

The group projection in the space formed by the first three axes confirmed that the first axis had a much greater discriminatory power than the two other, because the separation of groups was in fact much clearer along this axis. It distinguished samples with a maceration time greater than or equal to 24 hours from samples with a maceration time shorter than 24 hours (Fig. 9).

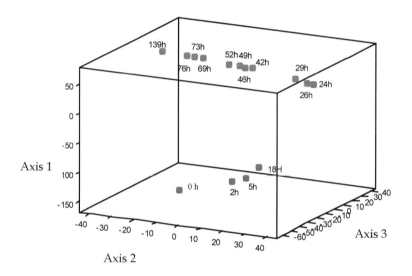

Fig. 9. Representation in the space formed by the first three discriminant axes of the factorial discriminant analysis (FDA) of the average scores of the olfactory fingerprint of the aromatic extract at 15 maceration times.

In addition, the second axis, which explained only 4.4 % of the variability, discriminated better between samples that have been macerated for at least 24 hours, and the third axis, which explained much less variability, better discriminated between samples that have undergone a maceration during less than 24 hours.

So, we can conclude that, according to the response of the sensors which depends on the affinity and concentration of the molecules in the head space, there was a significant difference between extracts from a maceration of less than one day and those from a maceration of one day or more.

3.6 Microbiological evolution of the plants extract

Microbiological monitoring of the aqueous extract during the maceration of plants focused on the enumeration of the microbial flora on the three culture media: MRS, PCA and PDA.

Fig. 10 shows the evolution of the logarithm of colony-forming units (cfu) per liter of the extract versus time of the plants maceration.

No bacterial growth was observed on MRS medium inoculated with the extract taken immediately after mixing MAPs with water, while the lactic microflora was dominant after 48 h and 72 h of maceration (10.14 and 11.03 Log(cfu/l) respectively). So, the development of the lactic acid bacteria was related to the maceration of MAPs, which has led to fermentation. Populations developed on the others two media (PCA and PDA) were respectively 9.15 Log(cfu/l) in 48 h and 9.92 Log(cfu/l) in 72 h for the mesophilic microflora and 9.60 Log(cfu/l) in 48 h and 10.22 Log(cfu/l) in 72 h for the fungal microflora.

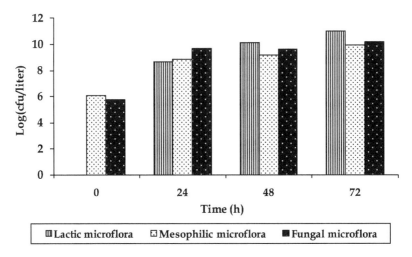

Fig. 10. Evolution of microbial flora (Log (cfu/liter)) during the maceration of medicinal and aromatic plants during 72 hours.

The entire fungal microflora developed in the extract was represented by yeasts. The absence of mold growth in the aromatic extract may be related to aromatic compounds known for their antifungal properties. Extended to three days, the antifungal effect of the aqueous extract has not changed, which may also inform about the release in water of all hydro-soluble active ingredients. It is clearly established that the antifungal activity of the aqueous extract of MAPs was due at least in part to the aromatic compounds they contain. These compounds, particularly those of thyme and oregano, had a marked influence on mold (Beraoud, 1990). The concentration of CO_2 in the jar could also inhibit the mold growth (Desgranges & Durand, 1990).

In addition, the chemical profile of an essential oil is very important due to its antifungal activity. Indeed, essential oil properties are due to all components it contains, especially to the chemical nature of its major components that would affect its fungicide activity in the following decreasing order: Aldehydes > Phenols > Ketones > Oxides > Hydrocarbons > Esters (where the ">" symbol means "more active than") (Beraoud, 1990). All of these chemical classes were found in our aromatic extract.

4. Conclusion

The study of physical, physico-chemical, biochemical and microbiological phenomena occurring during maceration of MAPs for the preparation of the aqueous aromatic extract used in manufacturing of *Tassabount* date juice, showed that the most of the physico-chemical changes took place after two days. The nature of biochemical reactions is fermentative, as indicated by a decrease in soluble solids including sugars, carbon dioxide and lactic acid production, with a predominance of the lactic acid bacteria flora.

According to the qualitative composition of aroma compounds of the MAPs extract, the best expression of the aromatic potential of MAPs was obtained after six hours of maceration, with the identification of 92 compounds. Terpenes and their alcoholic and ketonic derivatives were dominant among the identified aroma compounds.

Global olfactory fingerprint data showed also a very good discrimination between extracts from a maceration of less than one day and extracts from a maceration of one day or more.

Based on the majority of physical, physico-chemical, biochemical and microbiological studied criteria, a period of two days can be considered as an optimum duration of MAPs maceration. The duration limits are the first and third days of maceration.

Moreover, given the presence of bioactive compounds such as carvacrol, α-pinene, thymol and phenolic compunds, aromatic extract has biological activities that are not only related to the aromatization or the preservation (antimicrobial and antioxidant activities), but they are also correlated with functional properties that are potentially useful for pharmaceutical and nutritional applications.

The determination of the antimicrobial activity of the MAPs aqueous extract would be interesting for it exploration in the preservation of *Tassabount* date juice. In addition, the presence of lactic acid in this extract could be exploited for the *Tassabount* stabilization. In fact, its combination with essential oils, as an alternative of chemical preservatives and heat treatment, could be a valuable additive to preserve the nutritional and sensory quality of juice and prevent microbial alteration.

5. Acknowledgment

We gratefully acknowledge the Program for Agricultural Research for Development *"Processing dates of low-commercial value into juice: Evaluation of the quality and storage stability"* (PRAD 05/03, France-Morocco,) and the Fruit Trees Project, Research Agreement (APP/IAV-INRA-ENA-ENFI), Research Axis: *"Processing and Valorization of products"*, Research theme: *"Valorization by processing dates of low-commercial value in the oases"* (MCA Morocco, USA-Morocco,) for financially supporting this research.

6. References

Akrout, A. (2004). Etude des huiles essentielles de quelques plantes pastorales de la région de Matmata (Tunisie), In: *Réhabilitation des pâturages et des parcours en milieux*

Physico-Chemical, Biochemical and Microbiological Phenomena of the Medicinal and
Aromatic Plants Extract Used in the Preparation of Tassabount Date Juice in Morocco

127

méditerranéens, A. Ferchichi, (comp.), Cahiers Options Méditerranéennes, No.62, pp. 289-292, CIHEAM-IAMZ, Zaragoza, Spain

AOAC, 1990. *Official Methods of Analysis*. Association of Official Analytical Chemists, 15th Ed., Washington D.C., USA.

Beraoud, L. (1990). *Effet de certaines épices et plantes aromatiques et de leurs extraits sur la croissance et l'alflatoxinogénèse d'Aspergillus parasiticus NRRL 2999*. Thesis, Mohammed V University, Rabat, Morocco

De Vincenzi, M.; Stammati, A.; De Vincenzi, A. & Silano, M. (2004). Safety data review. Constituents of aromatic plants: carvacrol. *Fitoterapia*, Vol.75, (July 2004), pp. 801-804, doi: 10.1016/j.fitote.2004.05.002

Desgranges, C. & Durand, A. (1990). Effect of pCO_2 on growth, conidiation, and enzyme production in solid-state culture on *Aspergillus niger* and *Trichoderma viride* TS. *Enzyme Microb. Technol.*, Vol.12, (July 1990), pp. 546-551

Dimitrijević, S.I.; Mihajlovski, K.R.; Antonović, D.G.; Milanović-Stevanović, M.R. & Mijin D.Ž. (2007). A study of the synergistic antilisterial effects of a sub-lethal dose of lactic acid and essential oils from *Thymus vulgaris* L., *Rosmarinus officinalis* L. and *Origanum vulgare* L. *Food Chemistry*, Vol.104, Issue 2, (2007), pp. 774-782, doi: 10.1016/j.foodchem.2006.12.028

Georgé, S.; Brat, P.; Alter, P. & Amiot, M.J. (2005). Rapid determination of polyphenols and vitamin C in plant derived products. *J. Agric. Food Chem*, Vol.53, (2005), pp. 1370-1373, 10.1021/jf048396b

Hammiche, V. & Merad, R. (November 1997). *Peganum harmala* L., Available from http://www.inchem.org/documents/pims/plant/pim402fr.htm

Harrak, H. (2007). *Archivage, analyse et amélioration du savoir-faire traditionnel oasien : Cas du jus de dattes*. Thesis, Hassan II Agricultural and Veterinary Institute, ISBN 9954-444-20-3, Rabat, Morocco

Harrak, H.; Lebrun, M.; Ismaïli Alaoui, M.M.; Senhaji, A.F. & Hamouda, A. (2009). Vers une valorisation du savoir-faire local des oasis : Cas du jus de dattes *Tassabount* au Maroc. *Fruits*, Vol.64, No.4, (2009), pp. 253-260, doi: 10.1051/fruits/2009019

Harrak, H.; Reynes, M.; Lebrun, M.; Hamouda, A. & Brat, P. (2005). Identification et comparaison des composés volatils des fruits de huit variétés de dattes marocaines. *Fruits*, Vol.60, No.4, (2005), pp. 267-278, doi: 10.1051/fruits/2005033

Lee, S.; Umano, K.; Shibamoto, T. & Lee, K. (2005). Identification of volatile components in basil (*Ocimum basilicum* L.) and thyme leaves (*Thymus vulgaris* L.) and their antioxidant properties. *Food Chemistry*, Vol.91, (2005), pp. 131-137, doi: 10.1016/j.foodchem.2004.05.056

Nkhili, E.; El Hajji, H.; Tomao, V.; Dangles, O.; El Boustani, E. & Chemat, F. (2007). *Extraction des antioxydants du thé vert consommé au Maroc : Caractérisation et analyse par CLHP-UV-SM*. Congrès International sur les Plantes Médicinales et Aromatiques, CO63, Faculty of medecine, Fez, Morocco, March 22-24, 2007

Sijelmassi, A. (1996). *Les plantes médicinales du Maroc*, Le Fennec, 9981-838-02-0, Casablanca, Morocco

Zirari, A.; Harrak, H.; Chetto, A.; Alaoui Rachidi, M. & Outlioua, K. (2003). *Réhabilitation de la diversité génétique du palmier dattier dans la palmeraie de Fezouata*. Workshop, RAB/98/G31-PNUD/FEM/IPGRI Project, Zagora, Marocco, April 21-26, 2003

Part 3

Morphogenesis and Genetics

6

The Secretory Glands of
Asphodelus aestivus Flower

Thomas Sawidis
Department of Botany,
University of Thessaloniki, Thessaloniki, Macedonia,
Greece

1. Introduction

Asphodelus aestivus Brot. (*A. microcarpus* Viv.), family of Asphodelaceae (Asparagales), is a perennial spring-flowering geophyte, widely distributed over the Mediterranean basin (Tutin et al., 1980). Its formations represent the last degradation stage of the Mediterranean type ecosystems. These ecosystems often referred to as *"asphodel deserts"* or *"asphodel semi-deserts"* result from drought, frequent fires, soil erosion and overgrazing (Naveh, 1973; Pantis & Margaris, 1988). It is found both in arid and semi-arid Mediterranean ecosystems (Margaris, 1984) and in certain regions of North Africa (Le Houerou, 1979). *A. aestivus*, as in the case of other geophytes (cryptophytes), has a considerable distribution, since it has become the dominant life form in many degraded Mediterranean ecosystems. It is a widespread and invasive weed of the calcareous soils of pastures and grasslands, on hill slopes interspaced among cultivated areas, particularly abundant along road sides. The ability of *A. aestivus,* a native floristic element, to spread and to dominate in all those areas over the Mediterranean region reflects its capacity to face not only the peculiarities of the Mediterranean climate, but also to resist these most common disturbances in its habitat (Pantis & Margaris, 1988).

A. aestivus has two major phenological phases within a year. An active one (autumn - late spring) from leaf emergence to the senescence of the above-ground structures (photosynthetic period) and an inactive (summer) phase (dormancy), which lasts until the leaves emerge (Pantis et al., 1994). In February – March, the over wintering root tubers develop flat leaves (40-90 cm long x 2-4 cm width) and flowering stalks (70-170 cm tall) from a shoot apex. Fruiting starts in early May coinciding with the onset of leaf senescence. The fleshy leaves are herbivore protected by steroid saponins (Dahlgren et al., 1985) and senesce in June, before fruit maturation. The root tubers show lateral growth and vegetative propagation is frequent in mature plants. However, most of the root tubers remain attached to the mother plant. *A. aestivus* is a sessile organism reproducing by means of root tubers as well as by seeds. These facts are of considerable importance as far as maintenance and even dominance of *A. aestivus* within degraded areas are concerned.

The overproduction of flowers allows the plant to compensate for environmental variations and provides maternal chance in selective abortion of fruits and seeds (Stephenson, 1981;

Sutherland, 1986; Lee, 1988; Ehrlen, 1991). It is proved that the percentage of flowering is linearly related to the availability of nutrients (Pantis, 1993). The entomophilous flower of *A. aestivus* secretes a considerable amount of nectar, which is involved in pollination. As in many other plants, nectar is used to reward insects, which in turn offer a beneficial relationship (Simpson, 1993). The main pollinators are bumblebees and honeybees. Knuth (1899) provided general information about the floral biology of *A. albus* and Daumann (1970) described the morphology of the nectary. The release of the nectar onto the nectary surface occurs with a variety of mechanisms. Nectar may diffuse through the thin secretory cell walls or may accumulate beneath the cuticle of the nectary cells until the cuticle ruptures by the foraging vector releasing the nectar (Sawidis et al., 1987; Sawidis et al., 1989; Fahn, 1990; Sawidis, 1991; Sawidis et al. 2008). The above-ground structures (inflorescence stalks and fleshy leaves) are completely dry by June, and only the root tubers survive the dry summer (dormancy).

Asphodel meadows had first been referred by Homer. According to his epics (Odyssey, XI, 539, 573, XXIV, 13), the souls of the dead arrived in underground meadows *"asphodelos leimon"* on which only asphodels bloomed. The numerous underground root tubers of *A. aestivus* are up to 12 cm long and 4 cm thick. They have unlimited growth upwards, while the lower part breaks down. The age of the living part of a root tuber can be determined by the number of thickenings. The root tubers of *A. aestivus*, dried and boiled in water, yield a mucilaginous matter which in some countries, when mixed with flour or potato make the asphodel bread. In Spain and other countries, they are used as cattle and especially as sheep fodder. In Persia, a strong glue is made from the root tubers, which first become dried, pulverized and then mixed with cold water. Under the term *"Tsirisse,"* the root tubers of *Asphodelus bulbosus*, were used in eastern countries as a mucilage and to adulterate powdered salep. The ultrastructure and function of *A. aestivus* root tubers have been studied in order to explain the abundance of this plant in the Mediterranean region (Sawidis et al., 2005).

Because of the importance of *A. aestivus* as a consistent component of the Mediterranean vegetation and its dominance over wide areas, the present study aimed to evaluate the combined role of flower secretory glands and the mechanisms that contribute to its remarkable distribution to the Mediterranean region. The ultimate goal is to determine the morphology, anatomy, fine structure and function of septal nectary, osmophores and obturator and to correlate these structures with the remarkable success on pollination mechanism, fruit setting and thus the abundance of *A. aestivus* in the Mediterranean region. Still, reports of secretory structures in *A. aestivus* and other geophytes and their secretion mechanism are fragmentary.

2. Materials and methods

One-year-old plants of *A. aestivus* were collected from a hill about 25 km southwest of Larissa, Thessaly, Central Greece. Asphodel semi-deserts in Thessaly (Fig. 1) occupy an area of about 10.000 ha which is gradually expanding due to overgrazing, frequent fires and soil erosion (Pantis & Margaris, 1988). Flower buds of different age were sampled and floral parts were fixed with 2.5% glutaraldehyde and 2% paraformaldehyde in 0.05 M cacodylate buffer for 3 h. After post-fixation in 2% osmium tetroxide and dehydration in an ethanol series, the tissue was embedded in Spurr's epoxy resin. Cross sections were obtained in a

Reichert-Jung Ultracut E ultramicrotome. Semi-thin sections of 0.5-1.0 μm thickness from resin embedded tissue were stained with 0.5 % toluidine blue in 5% borax for preliminary light microscope (LM) observations. For qualitative detection of lignin in cell walls the cationic dye safranine O (1% aqueous), that show an affinity for lignin, was used.

For transmission electron microscopy (TEM) ultrathin sections (0.08 μm) were stained with 2 % (w/v) uranyl acetate followed by 2 % (w/v) lead citrate. Ultrastructural observations were carried out a Zeiss 9 S-2 and a Jeol GEM 1011 TEM. For scanning electron microscopy (SEM), specimens were fixed in 4 % glutaraldehyde cacodylate-buffered (pH 7.0) for 4 hours in room temperature without any osmium post fixation. After dehydration in an ethanol series (10 –100%), specimens were critical-point dried with liquid CO_2 as an intermediate and coated with gold in a CS 100 Sputter Coater. Observations were made using a BS-340 Tesla scanning electron microscope at various accelerating potentials. For polysaccharide staining, semithin sections of fixed or fresh material were treated with the periodic acid-Schiff's reagent (PAS) according to Nevalainen et al. (1972). For electron microscopic examination of polysaccharides ultrathin sections, collected on gold grids, were treated with periodic acid-thiosemicarbazide silver proteinate (PA-TCH-SP), according to Thiery (1967), following a procedure outlined by Roland (1978).

3. Results

3.1 Floral morphology

The flowers of *A. aestivus* are joined together in multi-branched pyramidal panicles. The number of flowers per inflorescence varies from 400 to 600 in April - May. In every cluster, the bottom flowers open first and then follow the above flowers. Every single flower remains in bloom for several days. The number of open flowers per inflorescence on the same day can be up to 30. The perianth of the actinomorphic flower consists of a distinct, heavily sclerefied calyx and corolla of six petals. The elongated white petals have a red/purple stripe through the centre. The centrally located gynoecium has a superior, spherical ovary with six distinct furrows (Fig. 2).

It is enclosed by a salmon/orange cap which is formed by six flaps, each coming from the base of each stamen. The light green ovary is well set off from the white, narrow and non-branched central style with slightly swollen stigma. The trilocular ovary has isomerous carpels, the lateral faces of which are united by fusion with one another (syncarpous gynoecium). In the middle of the ovary a hole connecting the septal slid and the outer space exists (Fig. 3). This hole (one per chamber) is the opening of the nectar and has the form of a groove which may be of different lengths. In younger flowers, the opening is small at pre-secretory stage (Fig. 4), whereas during the anthesis it grows much longer (Fig. 5). The epidermal cells of the ovary surface are nearby polygonal in outline having a strong relief (Fig. 6). Stomata, however, in the ovary epidermis are absent, probably due to the small dimensions of the organ. The six fertile androecial members of the perianth (stamens) are organized in to two whorls (3+3) and are free of each other. The outer line of stamens is slightly shorter than the inner one. The bases of the yellowish green filaments are wide and coalesce to form a cavity where nectar accumulates. The six conspicuously long stamens have orange anthers which produce large amounts of orange pollen. Stamens accrete with the basal part of the abaxial surface of filaments with the tepals of perianth.

Fig. 1. Asphodel semi-desert in Thessaly, Greece.

Fig. 2. Open flowers of *Asphodelus aestivus* consisting of six elongated petals. The flowers above bloom one by one.

3.2 The nectary

In cross section of the syncarpous gynoecium, the three carpels appear separated from each other by distinct septal slits. In these slits the tripartite floral nectary situated in the lower and middle part of the ovary (Fig. 7). The three septal slits proceed downwards entering the ascidiate zone of the carpels. The cavities are lined with secretory tissue which created

Fig. 3. Ovary formed by fusion of three carpels (syncarpous gynoecium). A narrow hole in the middle of the ovary (arrow) leading outwards the septal slid, X100.

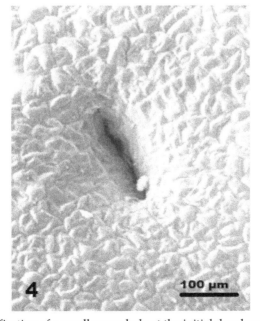

Fig. 4. Higher magnification of a small ovary hole at the initial development stages.

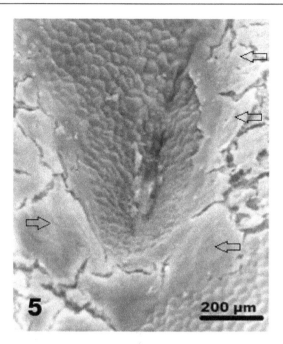

Fig. 5. Large nectary outlet at post secretory stage. Remnants of nectar are visible (arrows).

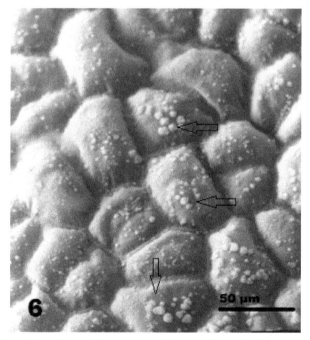

Fig. 6. Ovary epidermis surface with remnants of nectar (arrows). Stomata are absent.

1-3 layers of the nectary epithelium.   Each layer of epithelium consists of palisade or polygonal cells closely adpressed to the next ones. Epithelium cells of septal nectary are elongated and have very thin walls (Fig. 8). In the epithelium layer, stomata are absent and the cuticle lining the secretory cells is uniformly thin and electron opaque. In the nectariferous slits, at the initial development stages (pre-anthesis) structures like cytoplasmatic remnants are observed (Fig. 9), which may evidence lysigenous ontogeny of septal slits. Later, during the secretory stage, the nectar secreted by the epithelial cells accumulates in a space between the fusioned ovary carpels (slits).

At the beginning of anthesis (pre-secretory stage), the nectariferous tissue is well differentiated. The epithelium cells are more deeply stained than the surrounding subglandular parenchyma cells, containing a large, centrally situated nucleus and numerous organelles within a granular cytoplasm (Figs. 7, 8). The variously shaped plastids contain osmiophilic stroma, peripherally located thylacoids and different sized starch grains. The vacuoles are poorly developed in contrast to the subglandular tissue cells. The reduction of vacuome and a considerable increase of ER in epithelial cells characterize this stage. During anthesis (secretory stage) ER cisternae are widespread in epithelial cells, occupying a considerable fraction of the cell volume. At the stage of maximal development, the secretory epithelial cells contain a great number of mitochondria and Golgi bodies. The ER is well developed and occurs as long strands of parallel cisternae (Fig. 11). Active cisternal profiles of ER, in close contact with plastids dominate the secretory epithelium (Fig. 12).

At this time, the cavities (slits) in the middle part of the septal region between two adjacent carpels are narrow and a flockular substance, possibly nectar, is observed in the septal slits (Fig. 10). In some places, this substance is surrounded by a thin layer of cuticle and separated cuticle fragments are visible inside the slit of the nectar. The nectar is released to the outside through the outlets (holes) in the middle of the ovary (Figs. 3-5). Nectar secretion leads to an expansion of the space between the septa. At the post-secretory stage (two days after anthesis), the secretory cells of the nectary appear to have large vacuoles occupying the greatest cell volume, many of which contain electron dense bodies. This stage is characterized by completely hydrolyzed starch, as well as disappearance of the amyloplasts and ER. Remnants of the nectar on the adjacent of the outlets surface of the ovary have been observed.

3.3 The sub-glandular tissue

The nectary is supported by the subglandular tissue, which consists of about 10 layers of relatively large, isodiametric parenchyma cells, with large vacuoles and nuclei, as well as many plastids. In the subglandular tissue, unmodified, thin-walled crystalliferous idioblasts containing raphide bundles are present among ordinary parenchymatic cells.   These crystals are developed and stored rather passively within larger vacuoles of the specialised idioblast with a well distinctive tonoplast. Raphides appear in packs and wide morphological variations are observed among the cross sections (Figs. 13-15). Some electron dense substances penetrate the raphide surface and raphide grooves.   Raphides form in bundles of narrow, elongated needle-shaped calcium oxalate crystals, usually of similar orientation, with pointed ends at maturity, in idioblast cells.

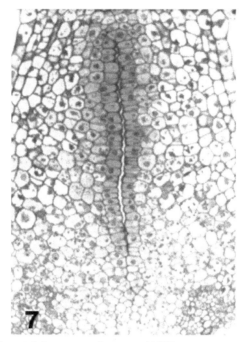

Fig. 7. An ovary carpel containing a septal nectary X 120.

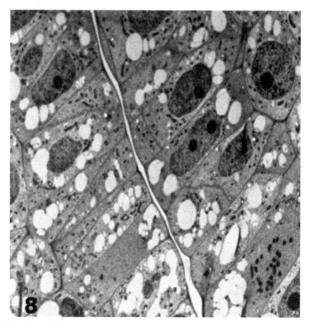

Fig. 8. Septal slit developed in nectary. The epidermal cells of the carpels contain dense cytoplasm and a large nuclei X 3.000.

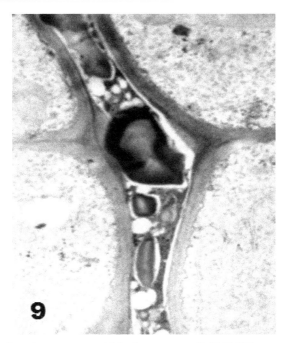

Fig. 9. Visible cytoplasmic remnants within the nectary slit X 20.000.

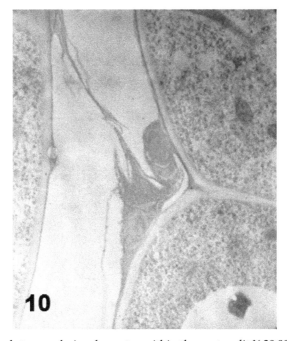

Fig. 10. Flockular substance, obviously nectar within the nectar slit X 20.000.

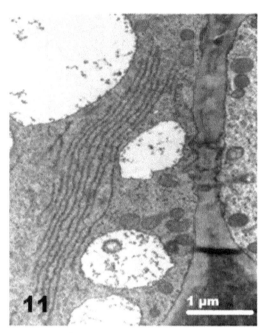

Fig. 11. ER occurring in strands of four or six cisternae at the lower part of the epidermal cell X 20.000. A number of plasmodesmata connecting the epidermal cells to subglandular tissue.

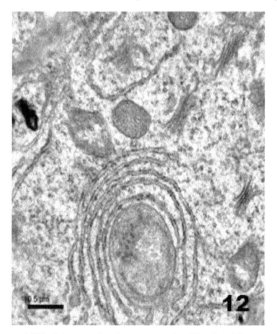

Fig. 12. Dictyosomes and ER cisternae in close contact with plastids.

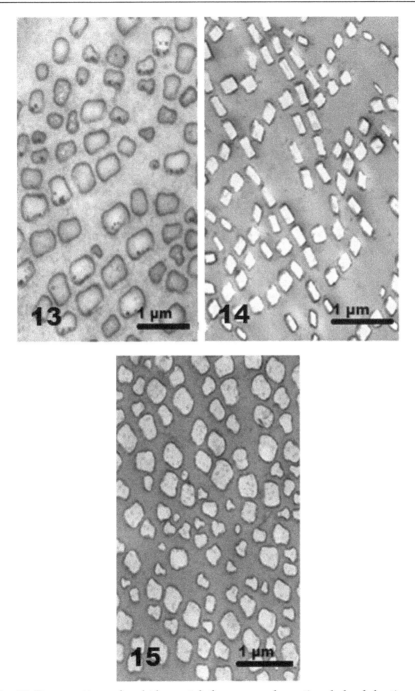

Fig. 13. - 15. Cross-sections of raphide crystals from parenchymatic subglandular tissue appeared in packs and wide morphological variations.

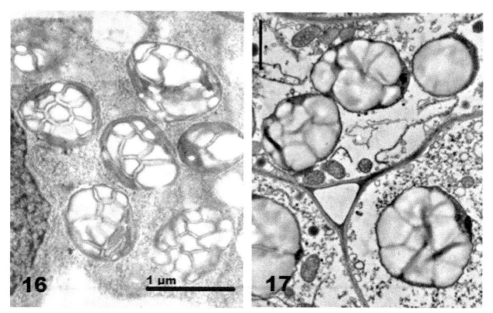

Fig. 16. - 17. Large starch grains in plastids of subglandular cells under the nectary (Fig. 16) and under the obturator (Fig. 17), at the initial developmental stages.

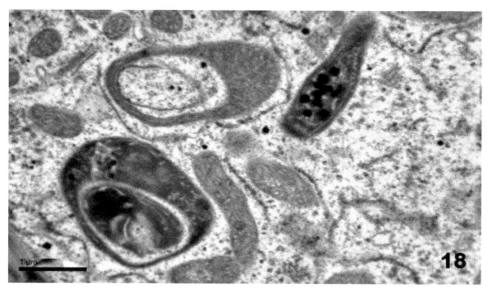

Fig. 18. Variously shaped plastids containing osmiophilic stroma without starch, after the secretory stage.

At the initial developmental stages of the nectar, enormous starch grains occur in cells of the subglandular tissue of various shapes and different sizes (Figs. 16, 17). Later, during nectar secretion, starch content drastically diminishes from the nectar epithelium towards the peripheral cells (Fig. 18). The use of Schiff's reagent allowed the exact location of starch in examined tissues (Figs. 19-21). At the beginning of the anthesis starch grains occur mainly in subglandular tissue cells. At this time, in epithelial, starch grains occurr sporadically and they are much smaller than in the subglandular parenchyma cells. Intercellular spaces of the subglandular tissue are very large and the parenchyma cells are connected to the nectary cells with plasmodesmata grouped in pit fields (Fig. 11). The nectary is supplied directly with well-developed vascular tissue, innervated exclusively by phloem strands. Xylem components are also noted in subglandular parenchyma.

3.4 The obturator

In the ovary of *A. aestivus* there are several ovules per carpel and placentation is axile. Between the lateral ovule and the ovary septa a central protrusion develops into a gland, the obturator. This is a prominent ovary wall outgrowth of placental origin, which lies in close contact with the micropyle of each ovule. The obturator is a secretory structure (mucilage gland), which gives a weak reaction (after employing the Schiff's reagent) at the initial developmental stages when starch the grains of the parenchyma cells are intensively red (Fig. 20). Later, during pollination, when starch has disappeared from the plastids, the presence of cells with a polysaccharidic content is more intense (Fig. 21). Starch stored within amyloplasts can be used both as a source of energy for highly metabolic processes and as a source of mucilage production. In the secretory cells of the obturator, numerous cisternal elements of ER in parallel arrangement are widespread occupying a considerable fraction of cell volume (Fig. 23). Numerous vacuoles containing amorphous electron dense material and mitochondria appear. Among ER cisternae, large vesicles with granular content occur. Golgi bodies are also prominent and consist of stacks of three to four cisternae.

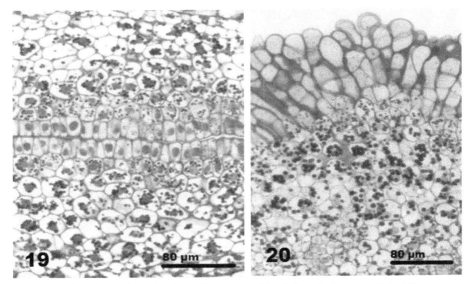

Fig. 19. - 20. Plastid starch grains from the nectary (fig. 19) and obturator (Fig. 20) subglandular tissue stained with the Schiff's reagent at the time before nectar secretion.

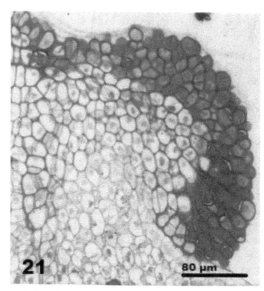

Fig. 21. Intense staining of the obturator papillae at the time after the nectar secretion. Plastid starch is completely absent from the subglandular tissue in comparison to Figs. 19 and 20.

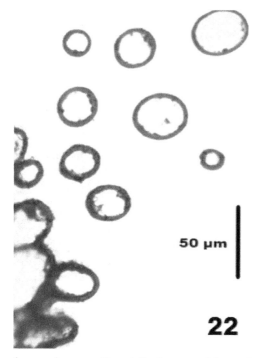

Fig. 22. Cross section of osmophore papillae at the bottom of the perianth tube. Cell walls stained red with the Schiff's reagent.

Mucilage is usually secreted by the Golgi apparatus and becomes processed by the ER (mucoprotein). The cells of the gland are internally surrounded by an extraplasmic space filled with the mucilage. Many dictyosomes with large vesicles at the ends of the cisternae and prominent ER elements reveal a secretory activity. Golgi bodies become associated with ER elements and bud off vesicles with a mucoproteinaceous content. The vesicles subsequently move to cell periphery, fuse with the plasmalemma and release their contents into the extraplasmic space between cell wall and plasmalemma.

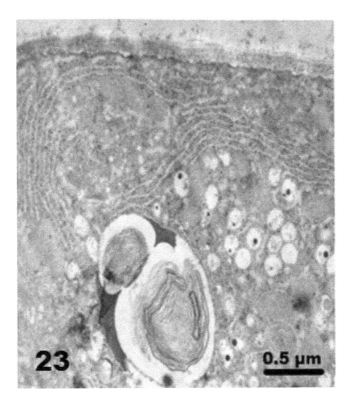

Fig. 23. Tip region of obturator papillae cell. ER cisternae occupying a considerable fraction of cell volume. Vacuoles of relative small size containing electron dence globules. Myelin like structures are also common.

This space progressively increases in volume towards the centre of the secretory cells at the expense of the protoplast (Fig. 24). In EM, mucilage can be detected after polysaccharide staining with the Thiery-reaction. A fine granular silver deposition reveales the fibrillar nature of the mucilage (Fig. 25). Silver deposits from the Thiery reaction are detected on the starch grains of amyloplasts in the subglandular parenchyma cells, whereas the rest of the amyloplast area appears completely negative to the reaction. In the cell wall silver grains are mostly deposited in the middle lamella consisting mainly of pectins (Fig. 26).

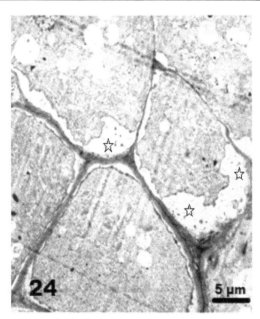

Fig. 24. Wall protuberances (asterisks) of the elongated papillae cells during the secretion process.

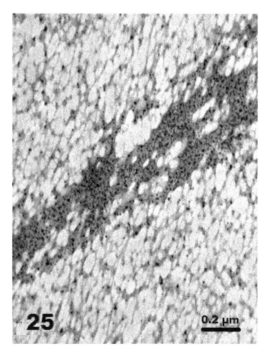

Fig. 25. Fibrillar mucilage material after polysaccharide staining with the Thiery-reaction.

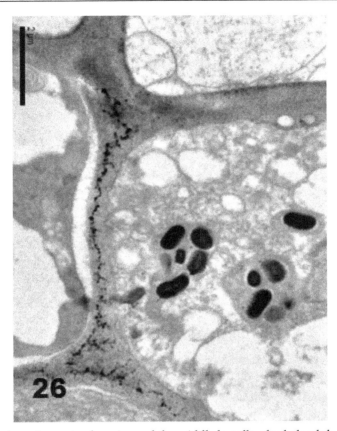

Fig. 26. Silver deposits on starch grains and the middle lamella of subglandular cells.

3.5 The osmophores

In *A. aestivus* flower the six, closely packed stamens surround the ovary in very close proximity to the nectar-releasing area. Their lower filament part is widened and flattened (Fig. 27), protecting the outlets of septal nectary. At this basal region of the stamen, the epidermal cells become larger and stretch outwards forming numerous papillose cells of various size and shape, the osmophores. The osmophores of *A. aestivus* are supported by subepidermal layers of parenchyma cells with large intercellular spaces (Fig. 27). In these subepidermal layers a well developed phloem is present. The centrally located vascular bundle is surrounded by closely layered parenchyma cells (Fig. 27). A feature also typical of osmophores is their intensive aeration by means of large intercellular spaces. In the cells of epithelium and subepidermal layers the presence of starch is also observed. The largest club-shape osmophores (approximately 170 μm), occur on the edges of the flattened part of filaments (Fig. 28). The outside, convex wall of the isodiametric papillose cells of the epidermis is considerably thick reaching on average 4.40 μm (Fig. 29).

PAS reaction reveals that the cell wall derives from cellulose (Fig. 22), while treating with safranine does not indicate any presence of lignin (Fig. 30). Osmophores possess a large

central vacuole whereas numerous small vacuoles are also present in the cytoplasm, which
enlarge during the emission of secretion and divide the cytoplasm into characteristic net-like
strips. On the longitudinal cross-sections the central part of the club shaped hairs and
papillae are filled usually by one vacuole, around which there is the cytoplasm which
creates a rather thin layer. In the cytoplasm, also small numerous vacuoles are observed
(Fig. 30), which gradually increase in size. Significantly enlarged nuclei (31 µm) in relation
to the nuclei in the parenchyma (12 µm) are usually located half way lengthwise in the club-
shaped papillae. In the cytoplasm, numerous, small plastids are observed. On the basal part
of filament osmophores are well developed in comparison to the above (Fig. 31), especially
on the filament edges. On their surface the cuticle reveals a striped ornamentation (Fig. 33).
On the top area of some of the papillose cells, round flattened areas are observed, which
probably mark breaking of the cuticle after emission of the previously accumulated elicitor
(Fig. 34). The cross-sections of the papillae have round or oval shapes (Fig. 22).

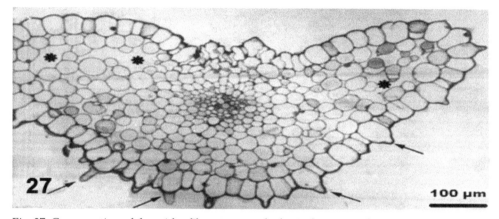

Fig. 27. Cross-section of the wider filament near the basis, bearing only one central vascular
bound. Papillose cells (arrows) and large intercellular spaces (asterisks) are visible.

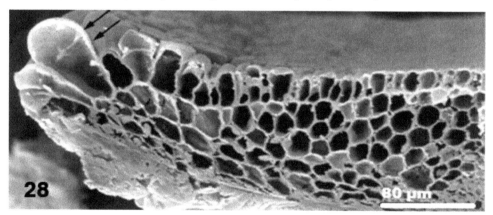

Fig. 28. SEM image of cross section at the lower part of filament forming at the edge
unicellular osmophore papillae (arrows).

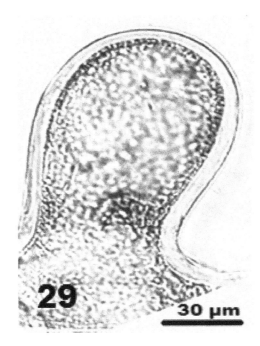

Fig. 29. Longitudinal section of osmophore papilla with thick cell wall.

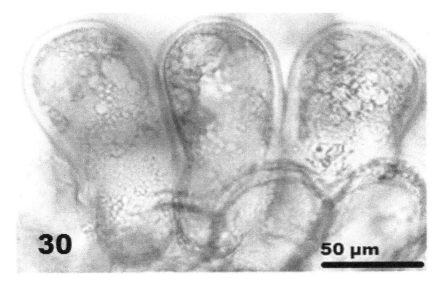

Fig. 30. Osmophores treated with safranin. Numerous small vacuoles in the cytoplasm.

Fig. 31. Lower part of a filament with densely developed osmophores.

Fig. 32. Osmophore papillae near the filament basis.

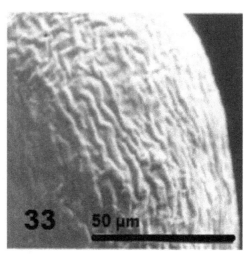

Fig. 33. Outer osmophore surface with irregular wrinkles.

Fig. 34. Circular traces on osmophore surface (asterisks) after emission of secretion.

4. Discussion

4.1 Nectar secretion and pollination

The septal nectary of *A. aestivus* formed by the incomplete fusion of the carpel flanks and lined by a secretory epithelium is the most common type of nectary in monocots (Vogel, 1998; Staufer et al., 2002). Tripartite nectaries are located in septal slits and their outlets have the form of elongate grooves which are situated above the middle part of the ovary. Septal slits in *A. aestivus* nectary are surrounded by 1-3 layers of epithelial cells, whereas in other plants, mostly one layer of the nectariferous tissue occurrs in the septal nectaries (Weberling, 1992; Smets et al., 2000). Gynopleural septal nectaries are considered the most advanced in the philogenetical development of plants. Many studies show that the migration of nectary glands proceeds in flowers from the perianth to internally located organs (Esau, 1965).

In *A. aestivus* nectaries, the epidermis of the septal slits participates in nectar secretion, since stomata in the nectary epidermis or on the ovary surface are absent. This fact agrees with the statement of Endress (1995) that nectarostomata are usually absent from monocotyledons. The side of the septal nectaries is diverse in the ovary in different representatives of monocotyledons. In some taxa, nectaries are situated at the base of the ovary with the outlet in the same region – *Johnsonia, Lomandra, Tofieldia* (Smets et al., 2000). In other genera, the nectar occupies the septal slits only in the upper part of the ovary with the outlet near the style base *Haemanthus, Astelia, Yucca* (Smets et al., 2000), *Haworthia* and *Gladiolus* (Weberling, 1992) and *Acidanthera* (Weryszko-Chmielewska et al., 2003). For *Allium* (Alliaceae) Maurizio and Grafl (1969) described the distribution of the opening of septal nectar at the half of the ovary height. This location is similar to the position of the outlets of the *A. aestivus* nectar (Weryszko et al. 2006).

Flowers of all species of *Asphodelus* are very attractive to pollinators since they produce large amounts of nectar (Cruden, 1977; Diaz Lifante, 1996). Some environmental factors, like drought, may affect the amount and viscosity of nectar influencing the visiting of pollinators (Harder, 1986). Changes of the nectary color provide insects with a signal that no reward is offered by a particular flower, thus forcing the insect to seek reward elsewhere. The maximum offer takes place during the morning, in a day of normal activity of insect visitors. The "large bee-dish shaped blossom" morphology of the *A. aestivus* flowers, as defined by Kugler (1977), allows the access of insects of a very wide range of size. The nectar is released to the outside through small holes on the ovary walls and accumulates in the base of the perianth tube. The accumulated fluid is retained until removed by a foraging insect. Ravena (2000) suggests that, in addition to the insect-like flowers, nectar derived from septal nectaries may be a further attraction to insect pollination. On the other hand, absence of nectar often indicates an alternative pollination mode (Rudal et al., 2002). Strong or distructive floral scents are associated with pollinator attraction (Hadacek and Weber, 2002; Effmert et al. 2005).

4.2 Obturator and fertilization

One of the secretory structures of the *A. aestivus* flower, connected with the process of fertilization, is the obturator. This is a mucilage gland, located between the lateral ovule and the ovary septa. The obturator is a placental protuberance at the ovary entrance connecting the transmitting tissue with the ovarian cavity (Tilton & Horner, 1980; Tilton et al., 1984; Herrero, 1992). The obturator acts a drawbridge, connecting the base of the style with the ovule. It contributes to the fertilization process by secreting different components involved in the growth of the pollen tubes just before they penetrate the micropyle (Tilton et al., 1984; Cheung, 1996). The obturator secretes a mucoproteinaceous product and gives a positive reaction after employing the Schiff's reagent. In spite of the early description of the obturator and its observation in a number of unrelated species very little is known about its function (Tilton & Horner, 1980; Herrero, 2000). Obturator invariably appears to support pollen tube growth on its way to the ovule, and it represents a further adaptation of a secretory placenta. By forming a little protuberance, it bridges the gap between the placenta and the ovule entrance, thereby facilitating the passage of the pollen tube in its journey towards the ovule, but only for a defined period, and at a specific stage of development (Herrero, 2003).

Pollen tubes first stop at the obturator, which lines the pollen tube pathway towards the ovule (Herrero, 2000). The pollen tubes are arrested at the obturator for a number of days until this structure enters a secretory phase accompanied by the vanishing of starch and accumulation of callose in the obturator (Arbeloa & Herrero, 1987). Pollen tubes must grow through the mucilage-filled intercellular spaces across the obturator before reaching the micropyle of the ovule (Webb & Williams, 1988; Clifford & Sedgley, 1993; Ciampolini et al., 1995; Weber & Frosch, 1995; Cheung, 1996). When the pollen tube arrives at the obturator the cells of the obturator surface are full of starch and devoid of secretion. Pollen tube growth is only resumed concomitantly with the production of this secretion providing nourishment for the growing pollen tubes. At the stage when the gland becomes functionally active in mucilage secretion entering the secretory phase the pollen tubes travel swiftly over this structure and enter the ovule (Gonzalez et al., 1996). The above pattern of secretion has been also observed in other mucilage glands (Lynch & Staechelin, 1995; Western et al., 2000).

4.3 Secretion mechanism

As to the manner of cellular secretion, the general literary consideration is that the type of granulocrine secretion is characterized by an abudance of active ER, mitochondria and Golgi. Eccrine secretion is characterized by relative few ER and Golgi but numerous plastids containing starch grains (Fahn, 1990). In the nectary of *A. aestivus,* the activity of the extensive ER, the number of mitochondria and the presence of vesicles in the Golgi cisternae suggest that granulocrine secretion is the mode of transport of nectar in this species (Rachmilevitz & Fahn, 1973; Fahn, 1979). Few starch grains are present in the nectariferous cells of *A. aestivus,* a fact implying that starch reserves contribute little to the nectar secreted. It is also suggested that the mode of nectar secretion from the nectariferous cells is granulocrine although further research is needed to confirm this hypothesis. Based on features of the ultrastructure of glandular cells it can be determined which pathway of nectar transport from inside to the outside of the nectariferous cells occurs.

The vascular tissue occurring in the subglandular parenchyma might serve the nectaries supplying them with carbohydrates. The bulk of nectar precursors come from the subglandular parenchymatic tissue and the phloem of the vascular bundles. Nectar secretion from the epithelial cells of the *A. aestivus* nectary slit takes place through a relatively thin cellulose wall. Nectar collects outside between the wall and the bulging layer of the cuticle, which then bursts, causing the release of nectar. After the nectar slit is filled up to about half of the height of the ovary, where the outlet of the nectary is situated, nectar flows outside and stops on expanded parts of filaments and perianth tepals. In *A. aestivus,* nectar production is supported by carbohydrate storing at the below-ground parts, since root tuber content of starch, lipids and soluble sugars varies considerably over the year (Sawidis et al., 2005).

Polysaccharides as well as total sugar contents are always higher in root tubers than in leaves. The highest values of soluble sugars appear in root tubers, in late spring-early summer. Since the *A. aestivus* root tuber biomass is 6- to 30-fold higher than that of the above ground plant parts (Pantis, 1993), this leads to the conclusion that the below ground part consists of a rather stable energy reserve under continuous replenishment. The greatest percentage of allocated biomass and nutrients is located in the tubers during the whole year. The nectar production is supported by stores in the below-ground parts. Hence, the changes of below- and above-ground biomass reflect adaptations, which synchronize the plant's phenological development with the seasonality of the Mediterranean climate. This seems to hold also in the case of *A. aestivus,* which has to withstand summer water stress in the semi-arid Mediterranean ecosystems. The deposition of nutrients allocated to root tubers of *A.aestivus* are highly variable, which is characteristic of plants living in unstable habitats (Muller, 1979), such as the asphodel semi-deserts. Changes in the contents of these compounds in the root tuber tissues suggest a massive translocation of soluble sugars over a year period (Meletiou-Christou et al., 1992).

4.4 Raphides and defence

Calcium oxalate needles in subglandular tissue are typical of raphide bundles found in other organs of *A. aestivus.* Isolated thin-walled crystalliferous idioblasts also occur in the root tubers where a greater variation of morphological differences of raphide cross sections is observed (Sawidis et al. 2005). Such morphological differences of raphides could potentially influence the degree of acridity. In parenchyma tissues the need for fight herbivores seem to be more vital. Irritation is both mechanical and chemical. When raphides come into contact with the tender tissues of worms and other herbivores these substances are injected, enter the wounds causing a traumatic injury and inflammation. This defense task, undertaken among others by raphides, seems to be vital to the parenchymatous subglandular tissue. The protection of *A. aestivus* raphides against herbivore attacks is reinforced by other cells containing defense compounds, such as alkaloids, found in other parenchymatic tissues supporting the nectary (Sawidis et al., 1998; Wittstock & Gershenzon, 2002).

Many plant cells contain crystalline inclusions of different chemical composition and shape. In monocotyledone, raphides are the most common type of calcium oxalate crystals, whereas in Asphodelaceae both raphides and styloids are often present at the same time (Prychid & Rudall, 1999). Raphides are needle-shaped calcium oxalate crystals that are produced by higher plants for defence, calcium storage and structural strength (Franceschi and Nakata 2005; Nakata 2003). They can occur in any plant organ or tissue, including stems, leaves, roots, tubers and seeds (Horner and Wagner 1995) and store calcium oxalates as metabolic waste or by-product of plant tissues. Accumulation of oxalic acid in tissues, which is not readily metabolized, may cause osmotic problems. Therefore, precipitation of calcium oxalate crystals seems to be an appropriate way for the plant to avoid these undesirable situations. The relationship between calcium ion absorption and oxalic acid synthesis in plants is most probably established for ionic balance in tissues to be maintained (Bosabalidis, 1987). On the other hand, the calcium content in both root tubers and subglandular tissue of nectaries may be viewed as an osmoregulatory mechanism during the secretion process (Evans et al., 1992).

4.5 Osmophores and scent dispersion

The inner side of the simple perianth of *A. aestivus* is a hot spot area containing nectaries and osmophores, which are variously formed and tend to occur in collateral pairs involved in reward production and the release of the odor (Teichert et al. 2009, Bolin et al. 2009). The papillae, located at the bottom of the perianth tube, belong to osmophores, which are floral organs for the manufacture, secretion and dispersion of scent (Vogel, 1990; Nilson, 2000). The numerous long papillae located at the upper part of the stamen basis form a sort of "barrier" that separates the nectar from the outer environment. This contributes to sealing off the nectar cavity and preventing it from evaporation (Loew & Kirschner 1911; Kugler, 1977). Nectar water loss leads to an exponential increase in viscosity, which makes the nectar collection by pollinators problematic (Manetas & Petropoulou, 2000). The fine structure and function of osmophores that emit fragrance has been studied by many researchers and the majority of the works deal with Orchidaceae, while the rest involve Asclepiadaceae, Aristolochiaceae, Araceae and Burmanniaceae (Weryszko-Chmielewska and Stpiczynska, 1995; Stpiczynska, 2001).

The ultrastructure of osmophores is involved in the mechanism of flower fragrance release, facilitating dispersal of the attractant chemicals (Paulus, 2006; Schiestl & Cozzolino, 2008). In the osmophores of many other plant species presence of starch grains is found. Starch is utilized as a source of both energy and carbon for the synthesis of volatile substances (Vogel 1990, de Melo 2010). Starch reserves in the tissues underlying the glandular epidermis, is a common characteristic of the osmophores (Vogel, 1990; Ascensao et al., 2005). Starch is the frequent energy reserve in osmophores, but lipids also occur. Numerous secretion vesicles with the lipid substances have been observed in the osmophore cells of many species by a number of researchers. Lipid substances, abundant in osmophores, have been presumed to be the physical counterpart of the secreted fragrance. This conjecture is based on the presumption that terpenes accumulate in that form (Curry, 1987; Pridgeon and Stern, 1983; Stern et al., 1987; Stpiczynska, 2001). The secretion product of osmophores is invisible, highly volatile and the amounts of each compound very low. The identified volatile compounds are mainly short-chained aliphatic aldehydes and alcohols (Vogel, 2000).

Scanning electron microscopy images have revealed a smooth and wax-powdered surface of the osmophores. The ablative wax particles which cover the osmophores inactivate the insects' tarsal pulvilli. This gliding device, in some species reinforced by zones of imbricate papillae, is irreversible, and no movements of floral parts allowing escape via the spathe mouth occur. In the subepidermal layers of the osmophores a well developed system of intercellular spaces is developed. Large air spaces in the parenchyma which facilitates their intensive aeration and a well developed phloem are typical features of osmophores (Vogel, 1990). The anatomical peculiarities of the scent gland and underlying tissues are consistent with the idea of a functional layering of the osmophore structure into storage, production and emission layers as found in many structured osmophores (Wiemer et al., 2008). The storage layer consists of parenchyma cells surrounding the vascular bundles, which are rich in starch grains that are consumed during flower anthesis. The production layer is constituted by the upper parenchyma and epidermal cells. There is an apparently intense flux of metabolites from this layer to the papilose cells, as suggested by conspicuous pit fields (Wiemer et al., 2008).

The function of osmophores has to be considered as a medium for flower scent release. The scent produced by osmophores, or other epidermal cells, directs the insects to the reproductive organs. Emission of volatile compounds by osmophores by cuticular diffusion processes has been observed in Orchidaceae before, such as in species of *Scaphosepalum*

(Pridgeon & Stern, 1985) and *Stanhopea* (Stern et al., 1987); or by cuticular pores in species of *Restrepia, Restrepiella* (Pridgeon & Stern 1983) and *Gymnadenia conopsea* (Stpiczynska, 2001). The emission of volatile compounds in these species of *Acianthera*, however, seems to be associated with the presence of stomata. Stomatal pores have frequently been observed on the surface of the nectaries that are involved in exogenous secretion, and Vogel (1990) suggests they could work as possible routes for volatile secretions. These compounds are probably volatilized by high daylight temperatures in the Mediterranean area and finally released in the outside environment through the cuticula. This hypothesis can be further supported by the liberation of these odors only during the hottest hours of the day (Borba and Semir, 2001; Borba and Semir 1998, De Melo 2010).

5. Conclusion

The secretory glands of *A. aestivus* flower, namely nectaries, obturator and osmophores play an important role in its strong reproductive performance. Their anatomical peculiarities are well adapted to a fluctuating environment and contribute to its successful sexual reproduction. These structural features combined with the ability of *A. aestivus* to avoid grazing and fires, may explain the species' frequent dominance in a wide area of arid environments, from the Mediterranean to the desert. The synchronized function of nectar secretion by the nectaries, with the nectar protection and scent emission by the osmophores and with the help of obdurator in the fertilization process bring forth the *A. aestivus* as the dominant life form in the degraded arid Mediterranean ecosystems.

6. Acknowledgment

I wish to thank the Hellenic Ministry of Education, Department of Inter-university Relations for financial support and Prof. Dr. Elzbieta Weryszko – Chmielewska for her valuable help in Scanning Electron Microscopy.

7. References

Arbeloa, A. & Herrero, M. 1987 The significance of the obturator in the control of pollen tube entry into the ovary in peach (*Prunus persica*). Annals of Botany, vol. 60, 681–685.

Ascensão, L., Francisco, A., Cotrim, H. & Pais, M. S. (2005). Comparative structure of the labellum in *Ophrys fusca* and *O. lutea* (Orchidaceae). American Journal of Botany, Vol.92, 1059–1067.

Bolin J. F., Maass E., & Musselman L. J. (2009). Pollination biology of Hydnora africana Thunb. (Hydnoraceae) in Namibia: brood-site mimicry with insect imprisonment. *International Journal of Plant Science, Vol.* 170, 157–163.

Borba, E. L. & Semir, J. (2001). Pollinator specificity and convergence in fly-pollinated Pleurothallis (Orchidaceae) species: a multiple population approach. Annals of Botany, Vol.88, 75-88.

Bosabalidis, A. M. (1987). Origin, ultrastructural estimation of possible manners of growth and non morphometric evaluation of calcium oxalate crystals in non-idioblastic parenchyma cells of *Tamarix aphylla* L. Journal of Submicroscopic Cytology, Vol.19, 423-432.

Cheung, Y. A. (1996). Pollen—pistil interactions during pollen-tube growth. Trends in Plant Science. Vol.1, 45-51.

Ciampolini, F., Faleri, C. & Cresti, M. (1995). Structural and cytochemical analysis of the stigma and style in *Tibouchina semidecandra* Cogn. (Melastomataceae). Annals of Botany, Vol.76, 421–427.

Clifford, S. C. & Sedgley, M. (1993). Pistil structure of *Banksia menziesii* R. Br. (Proteaceae) in relation to fertility. Australian Journal of Botany, Vol.41, 481–490.

Cruden, R. W. (1997). Pollen-ovule ratios: a conservative indicator of breeding systems in flowering plants. Evolution Vol.31, 32-46.

Dahlgren, R. M. T., Clifford, H. T. & Yeo, P. F. (1985). The families of the monocotyledons. Springer Verlag, Berlin.

Daumann, E. (1970). Das Blütennektarium der Monocotyledonen unter besonderer Berücksichtigung seiner systematischen und phylogenetischen Bedeutung. Feddes Repertorium, Vol.80, 463-590.

De Melo, M. C., Leite Borba, E. L. & Paiva, E. A. S. (2010). Morphological and histological characterization of the osmophores and nectaries of four species of *Acianthera* (Orchidaceae: Pleurothallidinae). Plant Systematic and Evolution, Vol.286,141–151.

Diaz Lifante, Z. (1996). Reproductive biology of *Asphodelus aestivus* (Asphodelaceae). Plant Systematic and Evolution, Vol.200, 177-191.

Effmert, U., Große, J., Röse, U. S. R., Ehrig, F., Kägi, R. & Piechulla, B. (2005). Volatile composition, emission pattern, and localization of floral scent emission in *Mirabilis jalapa* (Nyctaginaceae) *American Journal of Botany*, Vol.92, 2-12.

Ehrlen, J. (1991). Why do plants produce surplus flowers? A reserve-ovary model. American Naturalist, Vol.138, 918-933.

Endress, P. K. (1995). Major evolutionary traits of monocot flowers. In: P.J. Rudall, P.J. Cribb, D.F. Cutler, & C.J. Humphries (Eds.), Monocotyledons: systematics and evolution (pp. 43–79). Kew: Royal Botanic Gardens.

Esau, K. (1965). Plant anatomy. John Wiley and Sons. New York, London, Sydney.

Evans, R. D., Black, R. A., Loescher, W. H. & Fellows, R. H. (1992). Osmotic relations of the drought-tolerant shrub *Artemisia tridentata* in response to water stress. Plant, Cell and Environment, Vol.15, 49-59.

Fahn, A. (1979). Secretory tissues in plants. Academic Press, London.

Fahn, A. (1990). Plant Anatomy, 4th ed., Pergamon Press, Oxford, 587 p.

Franceschi, V. R. & Nakata, P. A. (2005). Calcium oxalate in plants: formation and function. Annual Review of Plant Biology, Vol. 56, 41-71.

Gonzalez, M. V., Coque, M., & Herrero, M. (1996). Pollen-pistil interaction in kiwifruit (Actinidia deliciosa; Actinidiaceae). American Journal of Botany, Vol.83, 148-154.

Hadacek, F. & Weber, M. (2002). Club-shaped organs as additional osmophores within the *Sauromatum* inflorescence: odour analysis, ultrastructural changes and pollination aspects. Plant Biology, Vol.4, 367-383.

Harder, L. D. (1986). Effects of nectar concentration and flower depth on flower handling efficiency of bumblebees.-Oecologia, Vol.69, 309-315.

Herrero, M. (1992). From pollination to fertilization in fruit trees. Plant Growth Regulation, Vol.11, 27-32.

Herrero, M. (2000). Changes in the Ovary Related to Pollen Tube Guidance. Annals of Botany, Vol.85, 179-85.

Herrero, M. (2003). Male and female synchrony and the regulation of mating in flowering plants. Philosophical Transactions of the Royal Society of London Vol.358, 1019-1024.

Horner, H. T., & Wagner, B. L. (1995). Calcium oxalate formation in higher plants. In S. R. Khan (ed.) Calcium Oxalate in Biological Systems, pp. 53-72. CRC Press, Boca Raton, Florida.

Knuth, P., (1899). Handbuch der Blütenbiologie. Bd. II, 2. Verlag von Wilhelm Engelmann, Leipzig, p. 490.

Kugler, H. (1977). Zur Bestäubung mediterraner Frühjahrsblüher. Flora, Vol.166, 43-64.

Le Houerou, H. N. (1979). North Africa. In: Goodall, D. E. and Perry R. A., eds., Arid Land Ecosystems. Vol.1, Cambridge University Press, Cambridge.

Lee, T. D. (1988). Patterns of fruit and seed production. In: Lovett-Doust J., Lovett-Doust L. (eds). Plant Reproductive Ecology. Oxford University Press, 179-202.

Loew, E. & Kirschner, O. (1911). Asphodelus L. In Kirschner, O., Loew, E., Schroter, C. (Eds). Lebensgeschichte der Blütenpflanzen Mitteleuropas 1(3), Ulmer, Stuttgart, pp. 296-303.

Lynch, M. A. & Staehelin, L. A. (1995). Immunocytochemical localization of cell-wall polysaccharides in the root-tip of Avena sativa. Protoplasma, Vol.188, 115–127.

Manetas, Y. & Petropoulou, Y. (2000). Nectar amount, pollinator visit duration and pollination success in the Mediterranean shrub Cistus creticus. Annals of Botany, Vol.86, 815-820.

Margaris, N. S. (1984). Desertification in Greece. Progress in Biometeorology, Vol.3, 120-128.

Maurizio, A. & Grafl, I. (1969). Das Trachtpflanzenbuch. Ehrenwirth Verlag, München.

Meletiou-Christou, M. S., Rhizopoulou, S. & Diamantoglou, S. (1992). Seasonal changes in carbohydrates, lipids and fatty acids of two Mediterranean dimorphic phrygana species. Biochemie und Physiologie der Pflanzen, Vol.188, 247-259.

Müller, R. N. (1979). Biomass accumulation and reproduction in Erythronium albidum. Bulletin Torrey Botanical Club, Vol.106, 276-283.

Nakata, P. A. (2003). Advances in our understanding of calcium oxalate crystal formation and function in plants. Plant Science, Vol.164, 901-909.

Naveh, Z. (1973). The Ecology of fire in Israel. In: Proc. 13th Tall Timber Fire Ecology Conference. Tallahassee. Florida, 139-170.

Nevalainen, J. J., Laitio, M., & Lindgren, I. (1972). Periodic acid-schiff (PAS) staining of Epon-embedded tissues for light microscopy. Acta Histochemica. Vol.42, 230-233.

Nilson, S. (2000). Fragrance glands (osmophores) in the family Oleaceae. In: Plant Systematics for the 21st Century (Eds. Nordenstam, G. El-Ghazaly, M. Kassas, Portland Press, London. pp. 305-320.

Pantis, J. & Margaris, N. S., (1988). Can systems dominated by asphodels be considered as semi-deserts? International Journal of Biométeorology, Vol.32, 87-91.

Pantis, J. (1993). Biomass and nutrient allocation patterns in the Mediterranean geophyte Asphodelus aestivus Brot. (Thessaly, Greece). Acta Ecologica, Vol.14, 489-500.

Pantis, J., Sgardelis, S. P. & Stamou, G. P. (1994). Asphodelus aestivus, an example of sychronization with the climate periodicity. International Journal of Biometeorology, Vol.32, 87-91.

Paulus, H. F. (2006). Deceived males – pollination biology of the Mediterranean orchid genus Ophrys. Journal Europäischer Orchidëen, Vol.38, 303–353.

Pridgeon, A. M. & Stern, W. L. (1983). Ultrastructure of osmophores in *Restrepia* (Orchidaceae). American Journal of Botany, Vol.70, 1233–1243.

Pridgeon, A. M. & Stern, W. L. (1985). Osmophores of *Scaphosepalum* (Orchidaceae). Botanical Gazette, Vol.146, 115–123.

Prychid, C. J. & Rudall, P. J. (1999). Calcium Oxalate Crystals in Monocotyledons: A Review of their Structure and Systematics. Annals of Botany, Vol.84, 725-739.

Rachmilevitz ,T. & Fahn, A. (1973). Ultrastructure of nectaries of *Vinca rosea* L., *Vinca major* L., and *Citrus sinensis* Osbeck cv. Valencia and its relation to the mechanism of nectar secretion. Annals of Botany, Vol.37, 1–9.

Ravena, P. (2000). The family Gilliesiaceae. Onira, Botanical Leaflets, Vol.4, 11-14.

Roland, J. C. (1978). General preparation and staining of thin sections. In: Electron microscopy and cytochemistry of plant cells. J. L. Hall (ed.). Elsevier, Amsterdam. p. 1-62.

Rudal, J. P., Bateman, M. R., Fay, F. M. & Eastman, A. (2002). Floral anatomy and systematics of Alliaceae with particular reference to *Gilliesia*, a perfumed insect mimic with strongly zygomorphic flowers. American Journal of Botany, Vol.89, 1867-1883.

Sawidis, T., Eleftheriou, E. P. & Tsekos, I. (1987). The floral nectaries of *Hibiscus rosa-sinensis* L. I. Development of the secretory hairs. Annals of Botany, Vol.59, 643-652.

Sawidis, T., Eleftheriou, E. P. & Tsekos, I. (1989). The floral nectaries of *Hibiscus rosa-sinensis* L. III. A morphometric and ultrastructural approach. Nordic Journal of Botany, Vol.9, 63-71.

Sawidis, T., (1991). A histochemical study of nectaries of *Hibiscus rosa-sinensis*. Journal of Experimental Botany, Vol.42, 1477-1487.

Sawidis, T., Kalyba, S. & Delivopoulos, S. (2005). The root-tuber anatomy of *Asphodelus aestivus*. Flora, Vol.200, 332-338.

Sawidis, T., Weryszko-Chmielewska, E., Anastasiou, V. & Bosabalidis, A. M. (2008). The secretory glands of *Asphodelus aestivus* flower. Biologia, Vol.63, 1118 – 1123.

Schiestl, F. & Cozzolino, S. (2008). Evolution of sexual mimicry in the orchid subtribe Orchidinae: the role of preadaptations in the attraction of male bees as pollinators. BMC Evolutionary Biology, Vol.8, 27 (10 pp).

Simpson, M. G. (1993). Septal nectary anatomy and phylogeny of the Haemodoraceae. Systematic Botany, Vol.18, 593–613.

Smets, E. F., Ronse Decraene, L.-P., Caris, P., & Rudall, P. J. (2000). Floral nectaries in monocotyledons: distribution and evolution. In: Wilson, K. L., Morrison, D. A., eds. Monocots: systematics and evolution. CSIRO: Collingwood, Victoria, 230–240.

Stauffer, W. F., Rutishauer, R. & Endress, K. P. (2002). Morphology and development of the female flowers in *Genoma interrupta* (Arecaceae). American Journal of Botany, Vol.89, 220-229.

Stephenson, A. G. (1981). Flower and fruit abortion: proximate causes and ultimate functions. Annual Review of Ecology and Systematics, Vol.12, 253-279.

Stern, W. L., Curry, K. J. & Pridgeon, A. M. (1987). Osmophores of *Stanhopea* (Orchidaceae). American Journal of Botany, Vol.74, 1323-1331.

Stpiczynska, M. (2001). Osmophores of the fragrant orchid *Gymnadenia conopsea* L. (Orchidaceae). Acta Societatis Botanicorum Poloniae, Vol.70, 91–96.

Sutherland, S. (1986). Patterns of fruit-set: what controls fruit-flower ratios in plants? Evolution 40, 117-128.

Teichert, H., Dötterl, S., Zimma, B., Ayasse, M., & Gottsberger, G. (2009). Perfume-collecting male euglossine bees as pollinators of a basal angiosperm: the case of *Unonopsis stipitata* (Annonaceae). Plant Biology, Vol.11, 29–37.

Thiery, J. P. (1967). Mise en evidence des polysaccharides sur coupes fines en microscopie electronique. Journal Microscopie, Vol.6, 987-1018.

Tilton, V. R. & Horner, H. T. (1980). Stigma, style and obturator of *Ornithogalum caudatum* (Liliaceae) and their function in the reproductive process. American Journal Botany, Vol.67, 1113–1131.

Tilton, V. R., Wilcox, L. W., Palmer, R. G. & Albertsen, M. C. (1984). Stigma, style, and obturator of soybean, *Glycine max* (L.) Merr. (Leguminosae) and their function in the reproductive process. American Journal Botany, Vol.71, 676–686.

Tutin, T. G., Heywood, V. H., Burges, N. A., Moore, D. M., Valentine, D. H., Walters, S. M., & Webb, D. A. (1980). Flora Europaea. Vol. V. Cambridge University Press. Cambidge.

Vogel, S. (1990). The role of scent glands in pollination: On the structure and function of osmophores. Smithsonian Institution Libraries and National Science Foundation. Washington, D. C., USA.

Vogel, S. (1998). Remarkable nectaries: structure, ecology, organophyletic perspectives. Nectar ducts. Flora, Vol.193, 113-131.

Vogel, S. (2000). A survey of the function of the lethal kettle traps of *Arisaema* (Araceae), with records of pollinating fungus gnats from Nepal. Botanical Journal of the Linnean Society, Vol.133, 61–100.

Webb, M. C. & Williams, E. G. (1988). The pollen tube pathway in the pistil of *Lycopersicon peruvianum*. Annals of Botany, Vol.61, 415–424.

Webber, M. & Frosch, A. (1995). The development of the transmitting tract in the pistil of *Hacquetia epipactis* (Apiaceae). International Journal of Plant Science, Vol.156, 615–621.

Weberling, F. (1992). Morphology of flowers and inflorescences. Cambridge, Cambridge University Press.

Weryszko Chmielewska, E. & Stpiczynska M. (1995). Osmophores of Amorphophallus rivieri Durieu (Araceae). Acta Societatis Botanicorum Poloniae, Vol.64, 121-129.

Weryszko-Chmielewska, E., Masierowska, M. & Laskowska, H. (2003). Budowa nektarnika acidantery dwubarwnej murielskiej (*Acidanthera bicolor* var. murielae Perry). Annales Universitatis Mariae Curie-Skłodowska, Sectio EEE, Horticultura, Vol.XIII, 123-127.

Weryszko-Chmielewska, E., Sawidis, T. & Piotrowska, K. (2006). Anatomy and ultrastructure of floral nectaries of *Asphodelus aestivus* Brot. (Asphodelaceae). Acta Agrobotanica, Vol.59, 29-42.

Weryszko-Chmielewska, E., Chwil, M. & Sawidis, T. (2007). Micromorphology and histochemical traits of stamina osmophores in *Asphodelus aestivus* Brot. flowers. Acta Agrobotanica, Vol.60, 13-23.

Western, T. L, Skinner, D. J. & Haughn, G. W. (2000). Differentiation of mucilage secretory cells of the *Arabidopsis thaliana* seed coat. Plant Physiology, Vol.122, 345-355.

Wiemer, A. P., Mori, M., Benitez-Vieyra, S., Cocucci, A. A., Raguso, R. A., & Sırsic, A. N. (2008). A simple floral fragrance and unusual osmophore structure in *Cyclopogon elatus* (Orchidaceae). Plant Biology, Vol.11, 506–514.

Wittstock, U. & Gerhenson, J. (2002). Constitutive plant toxins and their role in defence against herbivores and pathogenes. Current Opinion in Plant Biology, Vol.5, 1-8.

7

Microsporogenesis, Pollen Mitosis and *In Vitro* Pollen Tube Growth in *Leucojum aestivum* (Amaryllidaceae)

Nuran Ekici[1,*] and Feruzan Dane[2]
[1]Department of Science Education, Faculty of Education, University of Trakya, Edirne,
[2]Department of Biology, Faculty of Sciences, University of Trakya, Edirne,
Turkey

1. Introduction

The full range of gene expression leading to male gamete formation in flowering plants begins with determination of the stamen whorl in flower development and ends with release of mature sperm into the embryo sac near the egg and central cell (Bhojwani & Soh, 2001). This study focused on events from meiosis to pollen tube growth. Microsporogenesis has proved to be a highly informative character at the family level in angiosperm systematics, especially in monocots (Furness & Rudall, 1999). The two primary types of microsporogenesis – simultaneous and successive – differ in the relative timing of meiosis II, though intermediate conditions have been reported in some species. In the successive type, a callose wall is deposited after Meiosis I, thus forming a distinct dyad stage before the onset of Meiosis II; the resulting tetrads are predominantly tetragonal, decussate or occasionally T-shaped. In the simultaneous type, the two meioses proceed without interruption; cytoplasmic cleavage and deposition of the callose walls occur subsequently. With this type, tetrad shape is typically tetrahedral, though variable, ranging from tetragonal to tetrahedral (symmetric or asymmetric) or rhomboidal (Nadot et al., 2006). Among angiosperms, microsporogenesis seems to be more diverse within early-divergent lineages, possibly because the mechanisms controlling the process are more labile in these groups (Furness et al., 2002; Sajo et al., 2009). The asymmetric division of the microspore apparently activates a divergence of cellular programs to be initiated in the generative and vegetative cells (Park et al., 1998). Generative cell divides to form two sperm. When pollen grains are released from the anther, they are either bicellular, with one vegetative cell and one generative cell, or tricellular, with one vegetative cell or two sperm cells, depending on timing of generative cell division (Bhojwani & Soh, 2001). Pollen the male gametophyte of higher plants, is a biological system playing a central role in sexual plant reproduction (Cresti et al., 1992). Upon emergence from pollen grain, the pollen tube has to traverse distance often thousands times the diameter of the pollen grain to deliver the male gametes to the embryo sac for fertilization. The pollen tube growth pathway begins with pollen germination *in vitro* or *in vivo* (Cheung, 1996).

* Corresponding Author

This study presents new observations on microsporogenesis and pollen mitosis and pollen tube growth *in vitro* in Amaryllidaceae, in the context of an examination of cytological characters aimed at better understanding of relationships within Amaryllidaceae and other species. Previous investigations on these aspects are limited in Amaryllidaceae. According to recently records, Amaryllidaceae family has 60 genera and 800 species (Watson & Dallwitz, 2005). In recent years, *Leucojum* genus is represented by only two species (http://www.amaryllidaceae.org/Leucojum/). These are *L. aestivum* and *L. vernum*. The species that is widespread in Turkey and Thrace is *L. aestivum* subsp. *Pulchellum* (Davis, 1984). Embryological characteristics belong to this family have been gained from studies done with limited numbers of species (Şahin, 1997; Dane, 1998) and several characteristics of this family are inconsistent. It is very important to know about embryological characteristics of *L. aestivum* in terms of phylogeny. We aimed better understanding of the taxonomical characteristics of Amaryllidaceae family with this study. *L. aestivum* is also an economically important plant for galanthamin production (Heinrich, 2004) that exhibits low seed production. Microsporogenesis, pollen mitosis and pollen tube growth were investigated to understand the cause of low sexual reproductivity.

2. Material and methods

In this study, anthers of *L. aestivum* in various lengths were used. Materials were collected from Edirne Tavuk Forest in 2004-2005. Anthers were embedded in parafine and stained in Delafield's hematoxylene (Johansen, 1940). Different phases of microsporogenesis were investigated by using 1% aceto-orcein with squash preperation method. Pollen viability is studied with anilin blue prepared in lacto-phenol (Jensen, 1962). For pollen germination, the pollen grains from dehisced anthers were incubated in liquid germination medium [10% sucrose ($C_{12}H_{22}O_{11}$), 0.01% boric acid (H_3BO_3), 0.03% calcium nitrate ($Ca(NO_3)_2 \cdot 4H_2O$)] in test tubes containing 2 ml of medium for 4h at room temperature. Samples were fixed with acid-alcohol mixture (1 glacial acetic acid : 3 ethyl alcohol) following inoculation. Measuring 30 randomly selected zones on slides for 3000 pollen grains determined the *in vitro* germination percentages. Pollen was considered as germinated when the pollen tube was equal to or longer than the diameter of the pollen grains. Pollen tubes were stained with lactophenol anilin-blue and aceto orcein (Ünal, 1986). Photographs were taken with the help of Olymphus photomicroscope.

3. Results

In this study, microsporogenesis and pollen mitosis and *in vitro* pollen tube growth are described in *Leucojum aestivum* L. *(Amaryllidaceae)*.

3.1 Microsporogenesis

The young anther is bilocular and tetrasporangiate; the wall comprises four layers; epidermis; fibrous endothecium; 1-2 middle layers which disappears early and a multinucleate glandular tapetum (Plate 1). Tapetum layer degenerates in mature anthers of *L. aestivum* (Plate 2). In most microspore mother cells, the course of meiosis is regular (Plate 3). Cytokinesis is of the succesive type. Isobilateral, decussat, linear, and T-shaped tetrads were seen (Plate 4). In some cells irregularities were observed (Plate 5, Table 1), including chromosome bridges and lagging chromosomes.

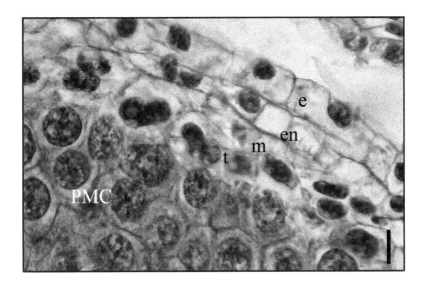

Plate 1. Cross section of young anther in *L. aestivum*; bar=20μm (e, epidermis; en, endothecium; m, middle layer; PMC, pollen mother cell; t, tapetum)

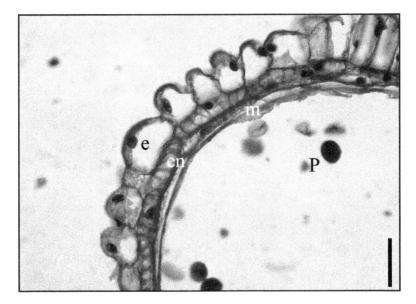

Plate 2. Cross section of mature anther in *L. aestivum*; bar=50μm (e, epidermis; en, endothecium; m, middle layer; P, pollen)

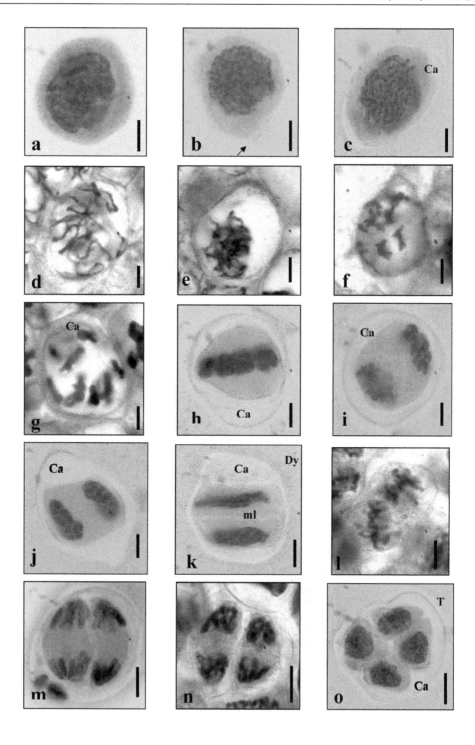

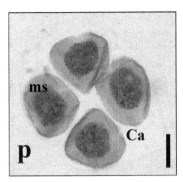

Plate 3. Successive type meiosis in pollen mother cells of *L. aestivum*. a, interphase; b,c leptotene, beginning of callose formation (arrow); d, zygotene; e, pachytene, chromosomes arranged like bouquet; f, diplotene; g, diakinesis; h, metaphase I; i, anaphase I; j, telophase I; k, dyad (interphase); k, prophase II; l, metaphase II; m, anaphase II; n, telophase II; o, tetrad; p, microspores. Bars=10μm (Dy, dyad; Ca, callose; ml, middle lamella; ms, microspore; T, tetrad)

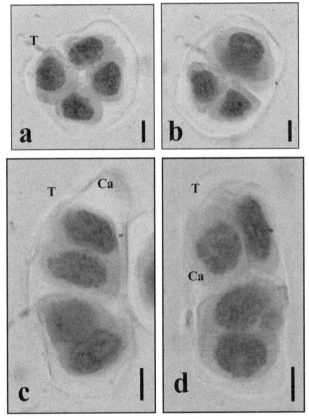

Plate 4. Tetrad types seen in *L. aestivum*; a, isobilateral; b, decussat; c, linear; d, T-shaped Bars=10μm (Ca, callose; T, tetrad)

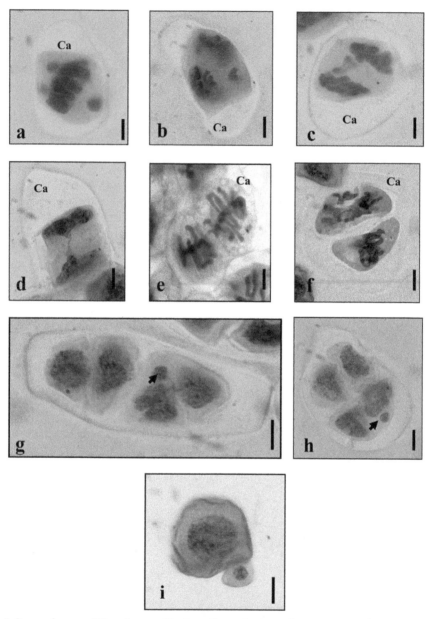

Plate 5. Some abnormalities observed in *L. aestivum* during microsporogenesis. a, chromosomes which were not arranged on equatorial plate in metaphase I; b, laggard chromosome in anaphase I; c, chromosome bridges in anaphase I; d, chromatid bridge in telophase I; e, laggard chromosome in metaphase II; f, irregularity on equatorial plate in metaphase II; g,h, micronuclei (arrows) in tetrad phase; i, normal and abnormal microspores including nucleus of various sizes. Bars=10μm (Ca, callose)

	Prophase I	Metaphase I	Anaphase I	Telophase I / Diad	Metaphase II	Anaphase II	Telophase II/ Tetrad	Total
Number of cell counted	228	112	28	176	100	68	308	1000
Number of normal cell	228	84	20	176	84	60	228	920
Number of abnormal cell	---	28	8	---	16	8	20	80
Percentage (%) of abnormal cell	0	25	28.5	0	20	11	6.4	8

Table 1. The percentage of abnormal cells observed in PMC of *L. aestivum* during meiosis.

3.2 Pollen mitosis

Microspores, which became free by being disrupted the callose in *L. aestivum* at the tetrad phase, were observed to be like a shell-shaped and their cell walls were seen wrinkled (Plate 6a). In the following phases, it was observed that microspores had been swollen that cell wall had flattened (Plate 6b) and vacuole had been formed on a pole (Plate 6c). Metaphase, anaphase and telophase were normal in pollen mitosis (Plate 6d-g). At the end of pollen mitosis, it was seen that nucleus nearby cell wall had formed generative cell and the other one had formed vegetative cell (Plate 6h,i). The pollen grains are 2 celled when shed (Plate 6j) but some abnormal pollen grains are seen (Plate 7a-c). Sterile pollen grains are also observed (Plate 8,9).

3.3 Pollen viability

The activity of pollens in *L. aestivum* was examined under light (Plate 8a,b) and flourescence microscope (Plate 9), they were usually stained well (Plate 8a). Pollens stained with aniline blue (Merck) and aniline blue (Plate 8b) prepared in Lactophenol solution were considered to be fertile. 17216 pollens were counted and the pollen sterility rate was found to be 1.7 %. The mean length and width values for pollen were determined 38.85±2.26µm and 24.36±2.1µm respectively by measuring the length of 100 pollens. Pollen grains are bilaterally symmetrical, monocolpate, prolate. The pollen grains are 2-celled when shed. The nucleus of the generative cell is long and lens-shaped and vegetative cell nucleus is lobulated.

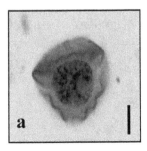
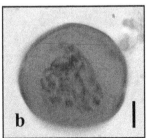
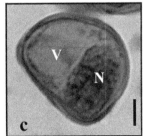

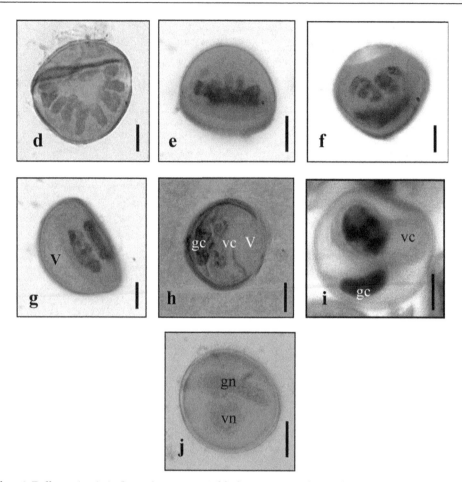

Plate 6. Pollen mitosis in *L. aestivum*. a, wrinkled microspore; b, swollen microspore; c, prophase in polarized microspore with vacuole; d,e metaphase; f, anaphase; g, telophase; h, pollen with vacuoles; i,j, mature pollen. Bars=10μm (gc, generative cell; gn, generative nucleus; N, nucleus; V, vacuole; vc, vegetative cell; vn, vegetative nucleus)

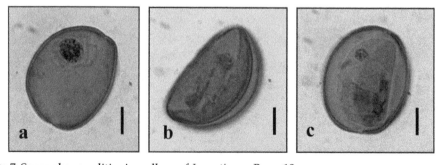

Plate 7. Some abnormalities in pollens of *L. aestivum*. Bars=10μm

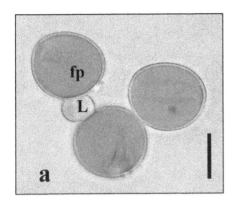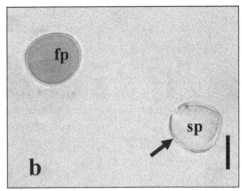

Plate 8. Mature pollen grains of *L. aestivum*; a, fertile pollens stained with aniline blue, b, fertile and sterile (arrow) pollens, Bars=20μm (sp, sterile pollen; fp, fertile pollen; L, lipid)

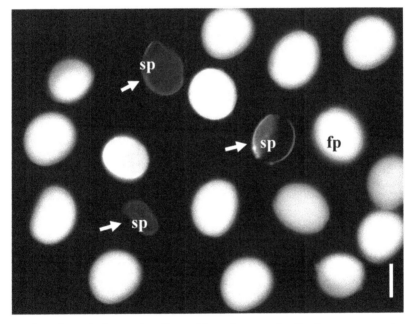

Plate 9. Fertile and sterile (arrows) pollens of *L. aestivum* under the flourescence microscope. Bar= 20μm (sp, sterile pollen; fp, fertile pollen)

3.4 Microgametophyte development

Microgametophyte development was examined in *in vitro* culture. Pollen grains obtained from the anthers in new blossoms (in the size of 15-20mm) of *L. aestivum* were germinated *in vitro* (Plate 10). 1000 pollens were counted and germination percentage of mature pollen grains was found to be 73%. Certain differences in terms of nucleus behaviors were found when germinated pollen tubes were cytological examined. In some pollen tubes it was observed that generative and vegetative nuclei did not move in tube and stayed in pollen (Plate 11a,b). In some others, however, it was observed that both nuclei move in pollen tube.

Normally, in some of these pollen tubes, vegetative nucleus moves ahead (Plate 10d), in some others generative nucleus moves ahead (Plate 11c,d). During *in vitro* pollen germination of *L. aestivum*, abnormal tubes were observed as well as normal ones. These abnormalities were weak development of some pollen tubes (Plate 11e,f) and swollen pollen tube tips (Plate 11g-k). Callose plug formation was not seen *in vitro* pollen tube growth in *L. aestivum*. *In vitro* pollen tube growth was also observed with fluorescence microscope (Plate 12).

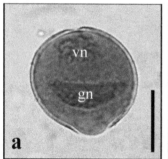
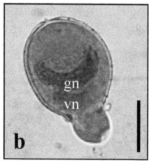
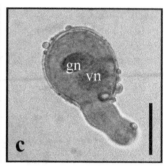

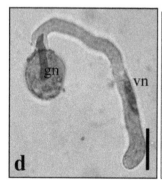
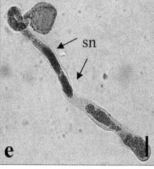

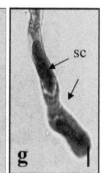

Plate 10. *In vitro* culture of microgametophyte development. a, pollen grain with two cells during germination Bar=20μm; b,c, vegetative and generative cell nuclei moving towards pollen tube, Bars=20μm; d, vegetative cell nucleus moving in prolonged tube Bar=20μm, e, sperm nuclei in pollen tube Bar=20μm; f,g, sperm cells at the tip of pollen tube, f, Bar=30μm; g, Bar=10μm (gn, generative nucleus; sc, sperm cell; sn sperm nucleus; vn, vegetative nucleus)

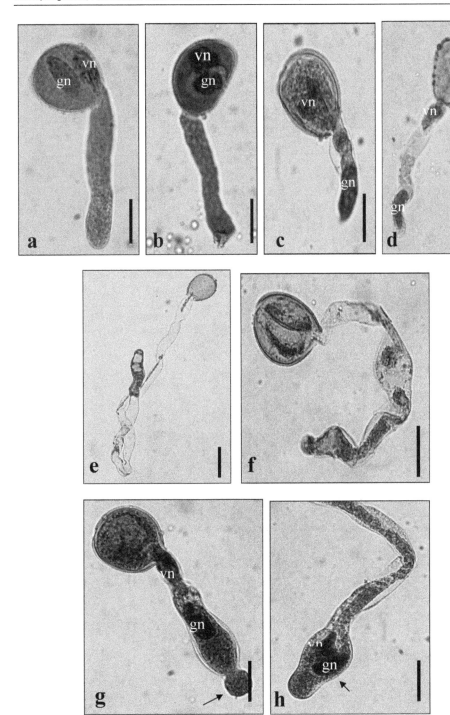

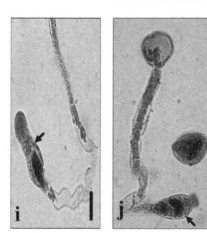

Plate 11. Some abnormalities of microgametophyte development during *in vitro* culture. a,b, pollen tubes without generative and vegetative nuclei in tube; c,d, pollen tubes in which generative nuclei went forward; e,f, weak development of pollen tubes; g-k, pollen tubes with swollen tips. a-d, f, g-k, Bar=20μm; e, Bar=30μm. (gn, generative nucleus; vn vegetative nucleus)

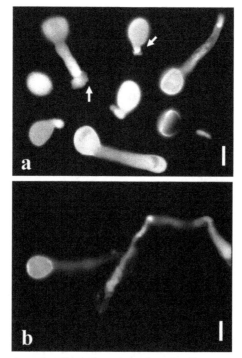

Plate 12. *In vitro* germination of pollen tubes viewed under fluorescence microscope a, developing pollen tubes and swollen tips of some pollen tubes (arrows); b, prolonged pollen tube; Bar=20μm.

4. Discussion

The anthers of *L. aestivum* are tetrasporangiate as in the other members of Amaryllidaceae. Its tapetum was with 1-2 cell lines and was glandular type like *Crinum* (Davis, 1966). Ameboid type of tapetum have been seen in *Galanthus* (Pankow 1958) that closely related with *Leucojum*. Microsporogenesis was observed to be generally regular in PMC of *L. aestivum*. Succesive type of cytokinesis is seen in PMCs during meiosis as in most of the monocotyledons (Furness & Rudall, 1999) and the other Amaryllidaceae members (Davis, 1966). Callose membrane, which began to form in the leptoten in PMCs of *L. aestivum*, dissolves just as in many angiosperms. On the other hand, in *Citrus limon* (Rutaceae) (Horner & Lernsten, 1971) callose deposition has taken place in interphase. Asynchronous meiosis in PMCs was seen in the anther locus of *L. aestivum*. On the other hand, synchronous microsporogenesis and simultaneous cytokinesis were seen in *Habenaria willd* (Orchidaceae) (Sharma & Vij, 1987). It was observed that PMCs at the top of some anthers in *L. aestivum* were in zygotene, whereas the ones at the bottom of the anther were in the pachytene. The phases at the top and the bottom of the anther were monitored to be far more different in advanced phases and the diakinesis stage was seen at the top, dyads were seen at the middle and tetrads were seen at the bottom of the anther. Furthermore, asynchronization was also determined among the anthers of the same bud. Microspore phase could be seen in one anther while tetrad phase was seen in another one. This was also identified in *Glicine max* (L.) Merr. (Fabaceae) (Albertsen & Palmer, 1979) and *Lilium longiflorum* L. (Liliaceae) (Shull & Menzel, 1977). Gradual transmitting of the nutrition from the bottom to the top can be the cause of asynchronization in the same locus of the anthers in *L. aestivum* and other plants. It was seen that meiosis I had usually been synchronous in the anther locus of *Coccina indica* aut. Non. (L.) Cogn., *Momordica charantia* Linn. and *Melothria maderaspatana* (L.) Cogn. (Cucurbitaceae) (Desphande et al., 1986); however, synchronization in meiosis II had become disrupted. This was explained by the existence of cytoplasmic channels in the early phases of meiosis I. It was seen in the ultrastructure of anther membrane of *L. aestivum* between the sporogeneous tissue cells. Then, in leptotene, cytoplasmic channels between PMCs have disappeared with the formation of callose membrane.

In this study, some irregularities were also seen although microsporogenesis in *L. aestivum* was generally regular. It was seen that the chromosomes had not been normally arranged in the equatorial plate in metaphase in some cells, lagging chromosomes were observed in anaphase I and telophase I and some chromosomes had formed choromosome bridges and there were micronuclei in tetrad and microspore stages. During anaphase II, micronuclei were formed by lagging chromosomes in *Gagea stipitata* (Liliaceae) (Koul et al., 1976), *Tulipa clusiana* (Liliaceae) (Wafai & Koul, 1982), *Allium textile* (Alliaceae) (Khaleel & Mitchell, 1982) and also *Bellevalia edirnensis* (Hyacinthaceae) (Dane, 2006). On the other hand, 1-8 micronuclei were seen in the tetrads of the artificial F_1 orchid hybrids. Darlington (1989) clarified that meiotic irregularities in both primitive and advanced groups of plants were resulted from the genetic defects and hybridization. In this study, a few meiotic irregularities were seen in *L. aestivum*. This could be caused by environmental factors as well as genetic instability. Lagging chromosomes did not reach the poles as a result of the absence of spindle fibers or irregular formation. This has a role in the micronuclei formation. In another case, on the other hand, PMCs were mitotically divided without reducing the chromosome number, chromosomes move towards the poles by seperating longitudinally and form diads in which chromosome number has not been reduced. Chromosome distribution is generally irregular between these dyads. Size of the nuclei is also different in dyads with unequal chromosome number. Some irregular chromosome-couplings leads to the delay in their movements towards the poles and causes them to be left in the middle of the cell. Some of these disappear and finally microspores with no equal size of nuclei are formed. However, some of them do not disappear and form micronuclei (Silva-Stort, 1984).

In *L. aestivum* like in the other Amaryllidaceae members, linear and T-shaped tetrads were rarely seen besides isobilateral, deccusate tetrads (Davis, 1966). On the other hand, in *Aristolochia elegans* L. (Aristolochiaceae) and *Sparganium erectum* L. (Sparganiaceae), 5 different types of tetrad were seen (Ünal, 2004). Isobilateral and tetrahedral tetrads were seen in *Tulipa clusiana* (Wafai & Koul, 1982) and *Gagea stipitata* (Koul et al., 1976) from Liliaceae; on the other hand, isobilateral tetrads were seen in *Bellevalia edirnensis* (Hyacinthaceae) (Dane, 2006).

Polarity was observed in pollen mitosis in *L. aestivum* as in *Sternbergia lutea* (L.) Ker-Gawler ex Sprengel (Dane, 1998) and *Bellevalia edirnensis* (Dane, 1999). During polarization, nucleus migrated to the proximal pole opposite to the aperture and a big vacuole was formed in the distal pole. This is peculiar to monocotyledones. It was determined that although nucleus migrated to the proximal pole in *Cypripedium fasciculatum* (Orchidaceae) (Brown & Lemmon, 1994) in pollen mitosis as in the other monocotyledones, nucleus migrates to the distal pole in *Phalenopsis* (Brown & Lemmon, 1992). During pollen mitosis, vacuolization (Ekici & Dane, 2004) has a role in the formation of polarization as well as nucleus migration (Brown & Lemmon, 1994; Ekici & Dane, 2004). According to Ünal (2004), the region, which nucleus migrates during the pollen mitosis, is identified the location of generative cell. This was also supported by our observations related to pollen mitosis of *L. aestivum*. As in the other Amaryllidaceae members (Davis, 1966), mature pollen grains are two celled when shed from the anther in *L. aestivum*. Generative cell nucleus is long and lens-shaped; on the other hand, the vegetative cell nucleus is lobed. Mature pollen grains two celled in *Sternbergia lutea* (Amaryllidaceae) (Dane, 1998), *Bellevalia edirnensis* (Hyacinthaceae) (Dane, 1999), *Tulipa clusiana* (Wafai & Koul, 1982) and *Gagea stipitata* (Koul et al., 1976), three celled in *Allium textile* (Alliaceae) (Khaleel & Mitchell, 1982) two or three celled in some of the species of the Araceae family (Davis, 1966) when shed. Mature pollen grains of *L. aestivum* are monocolpate as in *Galanthus ikariae* Baker and *G. rizehensis* Stern (Şahin, 1997). Pollens of *L. aestivum* and *G. rizehensis* are prolate; however, pollens of *G. ikariae* are subprolate.

Meiotic irregularities in sporogeneous cells cause spores to lose their viability. These spores cannot grow rapidly and absorb the nutrients normally supplied by the tapetum. As a result of this, sterile pollen grains are formed. Furthermore, a small ratio of pollen loose has been seen in the fertile plants. This ratio is approximately 15% and could vary between 2% and 20% (Zenkteller, 1962). On the other hand, in *L. aestivum* this ratio was determined as 1.77 %. According to Horner (1977), one of the reasons for pollen sterility is that nutrients cannot be transmitted into the microspores from the middle layer due to degeneration of the tapetal cells.

The tapetum is considered the source of callose so that abnormal development of the tapetum in the PMCs can cause failure of microspore release from tetrads and, thus, further development (Horner, 1977). In *L. aestivum*, the tapetum is normally developped. The cause of pollen sterility is meiotic irregularities.

Mature pollen grains of *L. aestivum* was germínated in *in vitro* medium. Germination percentage of mature pollen grains was identified as 73 %. During the formation of pollen tube, some differences in the behavior of the tube nucleus were observed. In two celled pollen grains, the cytoplasm filled the grain belongs to the vegetative cell and this also forms the cytoplasm in the pollen tube. Therefore, vegetative cell is called tube cell and its nucleus is named as tube nucleus (Ünal, 2004). When the pollen tube formed, vegetative nucleus usually firstly moves from the pollen grain. Even though This was observed in many angiosperms, it was seen that generative nucleus has moved first in some pollen tubes while vegetative nucleus moves first in some pollen tubes of *Triticum aestivum* L. (Chandra and Bhatnagar, 1974) and *Vicia* species (Dane and Meriç, 1999). During the formation of some pollen tubes, on the other hand, it was observed that both cell nuclei stayed in the pollen without moving along the pollen tube. The differences in nucleus behavior observed in pollen tube cells were also identified in *L. aestivum*. During *in vitro* pollen germination of *L. aestivum,* some abnormal tubes were seen besides normal tubes as in *Vicia galileae* Plitm. & Zoh. (Fabaceae) (Dane and Meriç, 1999). It was observed that some pollen tubes were developed less and some of the had swollen tube tips. This has been caused by the abnormalities in the nuclei of some microspores (Ünal, 2004).

5. Conclusion

Some cytological and embryological characteristics of *L. aestivum* are studied for the first time. Some irregularities were determined in the development of the male gametophyte but these irregularities do not affect fertility of plant. And also *Leucojum*'s tapetum is different from *Galanthus* the sister genus. The data that we gained from this study may contribute to embryological characteristics which were used in taxonomy of Amaryllidaceae (Meerow et al., 1999) family.

6. Acknowledgement

This study is a part of Nuran Ekici's PhD thesis and it is supported by Trakya University Scientific Research Projects with the project TUBAP-723.

7. References

Albertsen, M.C. & Palmer, R.G. (1979). A comperative light and electron-microscopic study of microsporogenesis in male sterile (MS1) and male fertile soybeans. *(Glycine max (L.) Merr.).* American Journal of Botany, Vol.66(3), pp.253-265, ISSN 0002-9122

Bhojwani, S.S. & Soh, W.Y. (2001). *Current Trends in the Embryology of Angiosperms,* Kluwer Academic Publishers, ISBN 0-7923-6888-6, Dordrecht, The Netherlands.

Brown, R.C. & Lemmon, B.E. (1992). Pollen development in Orchids 4. Cytoskeleton and Ultrastructure of the unequal pollen mitosis in *Phalaenopsis. Protoplasma*, Vol.167, pp.183-192, ISSN 0033-183X

Brown, R.C. & Lemmon, B.E. (1994). Pollen mitosis in the slipper orchid *Cypripedium fasciculatum. Sexual Plant Reproduction*, Vol.7, pp.87-94, ISSN 0934-0882

Chandra, S. & Bhatnagar, P. (1974). Reproductive Biology of *Triticum*. II. Pollen germination, pollen tube growth and its entry into the ovule. *Phytomorphology*, Vol.24(3&4), pp.211-217, ISSN 0031-9449

Cheung, A.Y. (1996). Pollen-pistil interactions during pollen tube growth, *Trends in Plant Science*, Vol.1, pp. 45-51, ISSN 1360-1385

Cresti, M., Blackmore, S. & Van Went, J.L. (1992). Atlas of Sexual Reproduction in Flowering Plants, Springer-Verlag, ISBN 3-540-54904-9, New York.

Dane, F. (1998). Mitotic divisions in root tip cells and pollen grains of *Sternbergia lutea* (L.) Ker-Gawl. Ex Sprengel (Amaryllidaceae). *Journal of Marmara for Pure and Applied Sciences*, Vol.14, pp.1-10. ISSN 1303-7412

Dane, F. (1999). Hekzaploid (2n=24) *Bellevalia edirnensis* Özhatay & Mathew'in polen mitozu ve polen morfolojisinin incelenmesi. *Turkish Journal of Biology*, Vol.23, pp.357-368, ISSN 1300-0152

Dane, F. (2006). Cytological and Embryological Studies of *Bellevalia edirnensis* Özhatay & Mathew (Hyacinthaceae). *Acta Botanica Hungarica*, September, Vol.57(3), pp.339-354, ISSN 0236-5383.

Dane, F. & Meriç, Ç. (1999). Reproductive biology of *Vicia* L. II. Cytological and cytoembryological studies on development of anther wall, microsporogenesis, pollen mitosis and male gametophyte in *V. galileae* Plitm. & Zoh. *Turkish Journal of Biology*, Vol.23, pp.269-281, ISSN 1300-0152

Darlington, C.D. (1989). Recent advances in cytology. (Genes, cells and organisms) Taylor & Francis ISBN 9780824013769, Mishawaka, USA

Davis, L.G., 1966, Systematic Embryology of the Angiosperms, Wiley, ISBN 101-224-154, New York, USA

Davis, P.H., 1984, Flora of Turkey and the East Aegean Islands. Edinburgh University Press, ISBN 0-7486-3777-X Edinburgh, United Kingdom

Deshpande, P.K., Bhuskute, S.M., Makde, K.H. (1986). Microsporogenesis and male gametophyte in some Cucurbitaceae. *Phytomorphology*, Vol.36(1,2), pp.145-150, ISSN 0031-9449

Ekici, N. & Dane, F. (2004). Polarity during sporogenesis and gametogenesis in plants. *Biologia, Bratislava* Vol.59(6), pp.687-696, ISSN 0006-3088

Furness, C.A., Rudall, P.J. (1999). Microsporogenesis in monocotyledons. *Annals of Botany*, Vol.84, pp.475-499, ISSN 0305-7364

Furness, C.A., Rudall, P.J. & Sampson, F.B. (2002). Evolution of microsporogenesis in angiosperms. *International Journal of Plant Sciences*, Vol.163, pp. 235-260, ISSN 1058-5893

Heinrich, M. (2004). Snowdrops: the heralds of spring and a modern drug for Alzheimer's disease. *The Pharmaceutical Journal*, Vol.273, pp.905-906, ISSN 0031-6873

Horner, Jr.H.T. (1977). A comparative light- and electron-microscopic study of microsporogenesis in male-fertile and cytoplasmic male-sterile sunflower (*Helianthus annuus*). *American Journal of Botany*, Vol.64(6), pp.745-759, ISSN 0002-9122

Horner, Jr.H.T. & Lernsten, N.R. (1971). Microsporogenesis in *Citrus limon* (Rutaceae). *American Journal of Botany*, Vol.58(1), pp.72-79, ISSN 0002-9122

Jensen, W.A. (1962). Botanical Histochemistry. W. H. Freeman and Company, ISBN 3-87429-257-6, San Francisco, USA.

Johansen, D. A., 1940, Plant Microtechnique. Mc Graw-Hill Book Company, ISBN 0-07-032540-5, New York, USA

Khaleel, T.F. & Mitchell, B.B. (1982). Cytoembryology of *Allium textile* Nels. and Macbr. *American Journal of Botany*, Vol.69(6), pp.950-956, ISSN 0002-9122

Koul, A.K., Wafai, B.A., Wakhlu, A.K. (1976). Studies on the genus *Gagea*. III. Sporogenesis. Early embryogeny and endosperm development in hexaploid *Gagea stipitata*. *Phytomorphology*, Vol.26(3), pp.255-263, ISSN 0031-9449

Meerow, A.W., Fay, M.F., Guy, C.L., Li, Q., Zaman, F.Q., Chase, M.W. (1999). Systematics of Amaryllidaceae based on cladistic analysis of plastid rbcL and trnL-F sequence data. *American Journal Of Botany*, Vol.86(9), pp.1325–1345, ISSN 0002-9122

Nadot, S.L., Forchoni, A. Penet, L., Sannier, J. & Ressayre, A. (2006). Links between early pollen development and aperture pattern in monocots. *Protoplasma*, Vol.228, pp.55-64.

Pankow, H. (1958). Über den pollenkitt bei *Galanthus nivalis*. *Flora* Vol.146, pp.240-253, ISSN 0367-2530

Park, S.K. Howden, R. & Twell D. (1998). The *Arabidopsis thaliana* gametophytic mutation gemini pollen 1 disrupts microspore polarity, division asymmetry and pollen cell fate, *Development*, Vol.125, pp. 3789-3799, ISSN 1011-6370

Sahin, N.F. (1997). An investigation on pollen morphology of *Galanthus ikariae* Baker and *Galanthus rizehensis* Stern (Amaryllidaceae). *Turkish Journal of Botany*, Vol.21, pp.305-307, ISSN 1300-008X

Sajo, M.G., Furness, C.A. & Rudall, P.J. (2009). Microsporogenesis is simultaneous in the early-divergent grass *Streptochaeta*, but successive in the closest grass relative, *Ecdeiocolea*. *Grana*, Vol.48, pp.27-37, ISSN 0017-3134

Sharma, M. & Vij, S.P. (1987). Embryological studies in Orchidaceae VI: *Habenaria willd*. *Phytomorphology*, Vol.37(4), pp.327-335, ISSN 0031-9449

Shull, J.K. & Menzel, M.Y. (1977). A study of reliability of synchrony in the development of pollen mother cells of *Lilium longiflorum* at the first meiotic prophase. *American Journal of Botany*, Vol.64(6), pp.670-679, ISSN 0002-9122

Silva-Stort, M.N. (1984). Sterility barriers of some artificial F1 Orchid hybrids. I. Microsporogenesis and pollen germination. *American Journal of Botany*, Vol.71(3), pp.309-318, ISSN 0002-9122.

Ünal, M. 1986. A comparative cytological study on compatible and incompatible polen tubes of *Petunia hybrida*, *Journal of Istanbul University, Faculty of Science*, Series B. Vol.51, pp.1-12, ISSN 0367-7753

Ünal, M. (2004). Plant (Angiosperm) Embryology. ISBN 975-400-040-9, İstanbul, TURKEY.

Wafai, B.A. & Koul, A.K., (1982). Analysis of breeding systems in *Tulipa*. II. sporogenesis, gametogenesis and embryogeny in tetraploid *T. clusiana*. *Phytomorphology*, Vol.32:(4), pp.289-301, ISSN 0031-9449

Watson, L., Dallwitz, M.J., 2005, The Families of Flowering Plants. http://delta-intkey.com/angio/www/amaryllidaceae.htm.

Zenkteler, M. (1962). Microsporogenesis and tapetal development in normal and male-sterile carrots (*Daucus carota*). *American Journal of Botany*, Vol.49(4), pp.341-348, ISSN 0002-9122

http://www.amaryllidaceae.org/Leucojum/

8

Ardisia crenata Complex (Primulaceae) Studies Using Morphological and Molecular Data

Wang Jun[1,2,3] and Xia Nian-He[1,*]

[1]Key Laboratory of Plant Resources Conservation and Sustainable Utilization,
South China Botanical Garden, Chinese Academy of Sciences, Guangzhou,
[2]Institute of Tropical Biosciences and Biotechnology,
Chinese Academy of Tropical Agricultural Sciences, Haikou,
[3]Graduate University of the Chinese Academy of Sciences, Beijing,
China

1. Introduction

Ardisia crenata Sims was a member of Myrsinaceae family in classical taxonomy view, but in the system of APG III (2009), it is included in the expanded family of Primulaceae and the primary Myrsinaceae family does not exist. This evergreen shrub is the most widely distributed species of *Ardisia*, occurring from Japan to Tibet, the Philippine Islands, and southern Asia where it is labelled as medicinal plant (Kobayashi & Mej'ia, 2005) and cultivated as a garden plant (Conover et al., 1989; Lee, 1998). Since *A. crenata* displays a high variability, its identification and species status frequently be confused. This complex includes four species and one variety (*Ardisia crenata* Sims, *A. hanceana* Mez, *A. lindleyana* D. Dietr., *A. linangensis* C. M. Hu, *A. crenata* var. *bicolor* (E. Walker) C. Y. Wu & C. Chen), They all belong to the subgenus *Crispardisia* of *Ardisia*. They have the same characters including inflorescences terminal, with leaf marginal nodules, 5 ovules in one series on the placenta. However, the five taxa also have some characters that could be indentified. There is some controversy between different researchers. Walker (1940) pointed out that *A. hanceana* is closely related to *A. crenata*, from which it may be distinguished by its larger flowers and usually by the lack of raised-punctate glands on the lower surface of the leaves. *A. hanceana*, *A. crenata* and *A. lindleyana* are very similar, the first one differ from *A. crenata* by the larger (6-7mm vs. 4-6mm) flowers, sepal ovate and differ from the last one by the marginal veins near the margin, more lateral veins (12-18 pairs vs. 8-12 pairs) (Chen, 1979). *A. linangensis* was first published by Hu (1992), he noted that this species differs from *A. hanceana* by the black-punctate flowers and by the not scalloped leaves and it is more closely allied to *A. tsangii* Walker, but can be easily distinguished by its glabrous and more corymbose inflorescence and by having more (3-8) leaves on the flowering branches. However, *A. tsangii* was treated as the synonym of *A. lindleyana* and *A. linangensis* was treated as the synonym of *A. crenata* in Flora of China (Chen & Pipoly, 1996). *Ardisia bicolor* was first

---

* Corresponding Author

published by Walker (1940), then it was dealt by Wu & Chen (1977) as a variety of *A. crenata* var. *bicolor*, they emphasized that the variety could be distinguished by the purple red of lower surface of leaves, peduncles, sepals and petals. Chen & Pipoly (1996) reduced the variety as the synonym of *A. crenata*. After we checked the specimens deposited in SCBG (Herbarium, Department of Taxonomy, South China Botanical Garden, Chinese Academy of Sciences), PE (Herbarium, Institute of Botany, Chinese Academy of Sciences), KUN (Herbarium, Kunming Institute of Botany, Chinese Academy of Sciences), IBK (Herbarium, Guangxi Institute of Botany), SYS (Herbarium, Life Science College, Sunyatsen University), HITBC (Herbarium, Xishuangbanna Tropical Botanical Garden, Chinese Academy of Sciences) and we found the identification of many specimens of them were incorrect. The five taxa are so similar in morphology, so it is necessary to clarify the relationships among them.

2. Materials and methods

2.1 Plant materials

Eighteen natural populations were sampled for molecular research. These representative populations were from Guangdong, Guangxi, Yunnan, Hainan Provinces. The geographical origins of accessions are given in Table 1. Voucher specimens were deposited in SCBG. Silica-gel dried samples of leaf tissue of each population were prepared for molecular analyses. We used 4 representational species belonging to other subgenera of the genus *Ardisia*, *Ardisia aberrans* (Pimelandra), *A. depressa* (Akosmos), *A. elliptica* (Tinus) and *A. japonica* (Bladhia) as an out-group. As for the morphological materials, we checked more than 2000 specimens in the Herbariums mentioned before and selected more than 570 specimens that have typical characters for morphology research.

2.2 DNA extraction

Genomic DNA was extracted following a modified 2×CTAB protocol (Doyle & Doyle, 1987) using samples of tissue cut from leaves. The total DNA of each sample was dissolved by 100μL Elution Buffer and diluted ten-fold before using for PCR. Total DNA was deposited at -20°C for long-stem storage. The quality of all DNA preparations was checked by agarose gel electrophoresis (1% w/v) in 0.5×TBE buffer containing 1 μg/mL of ethidium bromide by comparison with a known mass standard.

2.3 PCR amplification and DNA sequencing

ITS can be used for phylogenetic of species, and we also tried to use cpDNA gene, but we found it difficult to solve the relationships among these taxa. The two primers ITS1a (5'-AGAAGTCGTAACAAGGTTTCCGTAGG -3') and ITS4 (5'-TCCTCCGCTTATTGATATGC-3') (White et al., 1990) used in this study were designed on the basis of the regions of GenBank. The ITS sequences on GenBank are less, such as *Ardisia crenata* (FJ482136, FJ482137, FJ482138, AF547796), *A. japonica* (FJ482143, FJ482144, FJ482145, FJ482146), and we compared the sequence of the former with our sequences and found they were almost the same, for unifying the length of ITS regions, so in this research we just used the sequence we obtained.

For ITS PCR amplifications were performed in a total volume of 30 μl of reaction buffer, 1.5mmol/L MgCl$_2$, 10μmol of each primer, 2.5mmol/L of each dNTP, 5U/μl of Taq DNA polymerase and 10ng/µl of template DNA. Reactions were performed in a Peltier Thermal Cycler (Bio-RAD DNAEngine) and programmed for an initial denaturation step (3 min at 94 °C) followed by 35 cycles of 45S at 94°C, 50S at 55°C, 1min at 72°C. The last cycle was followed by a final incubation of 10 min at 72°C. Subsequently, 3µl of each amplification mixture was analyzed by agarose gel (1% w/v) electrophoresis in TBE buffer containing 1µg/mL ethidium bromide. The PCR reacions were purified from excess salts and primer using the Qiagen QLAquick PCR purification Kit. Automated DNA sequencing was performed directly from the purified PCR products using ABI 3730 DNA sequencer (Applied Biosystems) by Shanghai Invitrogen biotechnology Co.Ltd. and Shanghai Biosune biotechnology Co.Ltd. All sequences of ITS were bi-directional sequenced and the region was not cloned.

2.4 Sequence alignment and analysis

DNA sequences and overlapping fragments were assembled and edited using SeqMan and checked for orthology to sequences of *Ardisia crenata* complex. The sequence boundaries between the two spacers (ITS1 and ITS2) and coding regions (5.8S) of nrDNA were determined by comparison with the *A. crenata* sequence (Hao et al., 2003).

Multiple alignments were automatically performed using CLUSTAL X 1.83 (Thomson et al., 1997) of DNA Star (DNASTAR Madison, WI), and then further examined and slightly modified manually.

Phylogenetic analyses for each matrix were carried out Bioctrl package using maximum parsimony (MP) and Bayesian inference (BI) methods in PAUP* 4.0b10 (Swofford, 2001) and MrBayes version 3.12 (Ronquist et al., 2003; Huelsenbeck & Ronquist, 2001). For MP analyses, heuristic searches were conducted with 1000 replicates of random addition, one tree held at each step during stepwise addition, tree-bisection-reconnection (TBR) branch swapping, MulTrees options on, and the steepest descent off, Gaps were treated as missing data, characters were equally weighted, and their states were unordered. Relative clade support was evaluated by the bootstrap analyses (Felsenstein, 1985). For Bayesian analyses were accomplished in MrBayes version 3.12 using the best-fit models upon Akaike information criterion (AIC; Akaike, 1974) by using Modeltest 3.7 (Posada & Crandall, 1998; Posada & Buckley, 2004). In Bayesian analyses, trees were generated by running four simultaneous Metropolis-coupled Monte Carlo Markov (MCMC) chains and sampling one tree every 1000 generations for 1,000,000 starting with a random tree. The posterior probability (PP) was used to estimate nodal robustness.

3. Results

3.1 Sequence characteristics

All the acquired sequences have been submitted to GenBank and can be retrieved using the numbers in Table 1. No evidence of paralogous sequences was found for ITS sequences, because all PCR products were resolved as a single band and no double peaks were encountered in sequencing. The ITS region of nrDNA comprising both ITS sequences (ITS1 and ITS2) and the 5.8S rDNA was amplified by PCR form all 18 taxa of the *A. crenata* complex and 4 samples of outgroup. The ITS aligned sequence data set was 681 bp in length, with 46 positions being variable and 33 parsimony-informative.

Name of species	Subgenus	Locality	Voucher	GB No. ITS
Ardisia crenata Sims	*Crispardisia*	Guangxi, China	J. Wang 200810	JN645183
		Guangxi, China	J. Wang 2007146	JN645181
		Yunnan, China	J. Wang 2007218	JN645182
		Guangdong, China	J. Wang 2007299	JN645180
A. crenata var. *bicolor* (E. Walker) C. Y. Wu & C. Chen	*Crispardisia*	Guangdong, China	J. Wang 2007121	JN645184
		Guangdong, China	J. Wang 2007123	JN645185
		Guangdong, China	J. Wang 2007293	JN645186
A. hanceana Mez	*Crispardisia*	Guangdong, China	J. Wang 200857	JN645190
		Hainan, China	J. Wang 200799	JN645187
		Yunnan, China	J. Wang 2007217	JN645188
A. lindleyana D. Dietr.	*Crispardisia*	Guangdong, China	J. Wang 2007296	JN645189
		Guangdong, China	J. Wang 200604	JN645191
		Guangdong, China	J. Wang 200858	JN645193
		Guangdong, China	J. Wang 2007112	JN645192
A. linangensis C. M. Hu	*Crispardisia*	Guangdong, China	J. Wang 2007119	JN645196
		Guangdong, China	J. Wang 200652	JN645194
		Guangdong, China	J. Wang 200653	JN645195
		Guangdong, China	J. Wang 2007215	JN645197
A. aberrans (E. Walker) C. Y. Wu & C. Chen	*Pimelandra*	Kachin, Myanmar	Xia et al. 381	JN645198
A. depressa C. B. Clarke	*Akosmos*	Guangxi, China	J. Wang 2007169	JN645199
A. elliptica Thunb.	*Tinus*	Guangdong ,China	J. Wang 2007301	JN645200
A. japonica (Thunb.) Bl.	*Bladhia*	Hunan, China	Y. Z. Chen 2007298	JN645201

Table 1. Origin of samples, voucher information, and GenBank database accession numbers of DNA sequences of *Ardisia crenata* complex.

3.2 ITS analysis

The consensus MP phylogenetic tree (L= 95, CI= 0.874, RI= 0.898) and the Bayesian tree derived from ITS/5.8S sequences was shown with bootstrap values in Fig. 1 and Fig. 2. The topology of the strict consensus tree and Bayesian tree are almost identical, just the support values are different, the latter were higher than the former.

All constructed ITS phylogenetic trees congruously suggested that the *A. crenata* complex was divided into three clades, which are strongly supported (MPBPs/BPP: 87/0.98 in clade A; MPBPs/BPP: 62/0.88 in clade B; MPBPs/BPP: 95/1.00 in clade C). Clade A and clade B are composed of all of the samples of *A. hanceana* and *A. lindleyana* separately, *A. crenata*, *A. crenata* var. *bicolor* and *A. linangensis* clustered together with high bootstrap value (MPBPs/BPP: 95/1.00), confirming very close relationships among them. All samples of *A. crenata* and two samples of *A. crenata* var. *bicolor* clustered in Clade D with the bootstrap value (MPBPs/BPP: 93/1.00). All samples of *A. linangensis* and one *A. crenata* var. *bicolor* were in multiple clades.

3.3 Morphological and anatomical research

After identifying more than 570 specimens of this complex and dissecting some specimens' flowers, we can easily distinguish *Ardisia lindleyana*, *A. hanceana* from *A. crenata*, *A. crenata* var. *bicolor*, *A. linangensis* in morphology. *A. crenata*, *A. crenata* var. *bicolor* and *A. linangensis* are very similar with each other and many characters of them have different degree of transition, but the typical characters can be easily identified. In this research, we select the typical characters used for summarization, the results are submitted below (Table 2).

Characters/ name of species	*Ardisia crenata*	*A. crenata* var. *bicolor*	*A. linangensis*	*A. hanceana*	*A. lindleyana*
Leaf shape	elliptic to elliptical lanceolate	elliptic to long oblong-elliptical lanceolate	long oblong-elliptical lanceolate or elliptic	elliptic or long oblong-elliptical lanceolate	oblong to elliptical lanceolate
Leaf margin	crisped or crenate	crisped, crenate or dentate	entire or dentate	crenate or entire	entire or dentate
Marginal punctate	obvious, irregular	obvious, uniserial or irregular	obvious, uniserial	without or unconspicuous	unconspicuous or irregular
Leaf margin nodules	between teeth, densely	between or on teeth, densely or sparsely	without or on teeth, sparsely	between teeth, sparsely	without or on teeth sparsely
Flower colour	white	purple to faint red	faint red	Faint red to purple	white
Lobe apex	acute	acute or obtuse	obtuse	acuminate	obtuse
Inflorescences	umbellate or cymose	umbellate or cymose	corymbose or umbellate	compound corymbose cyme	umbel
Flower	4-6mm	4-6mm	5-6mm	6-7mm	ca. 5mm
Flowering branches	4-16cm	2.5-12cm	9-24cm	8-24cm	3-11cm
Leaves of Flowering branches	0-5 pieces	2-5 pieces	3-8 pieces	5-16 pieces	1-3 pieces
Pedicel length	0.5-1.5cm	0.5-1.5cm	1-3cm	1-2.5cm	0.5-1.5cm
Sepal shape	triangular ovate	triangular ovate	triangular or elliptical ovate	ovate	Obtuse triangular ovate
Sepal and pedicel indumentums	without	without	without	without	with
Sepal punctate	yellow-brown	black	black	almost without punctate	black
Fruit diameter	5-6mm	5-6mm	7-8mm	7-9mm	4-6mm

Table 2. Morphological and Anatomical comparison of *Ardisia crenata* Complex

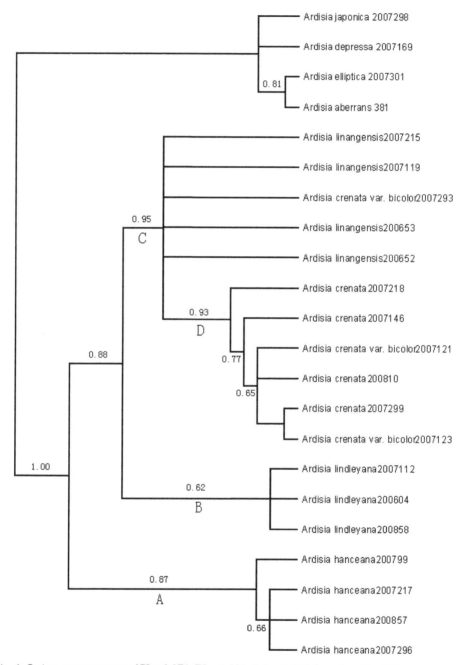

Fig. 1. Strict consensus tree (CI = 0.874, RI = 0.898, RC = 0.126) based on the ITS sequence data for 22 taxa of the *A. crenata* complex, and the number above branches indicated bootstrap values above 50%

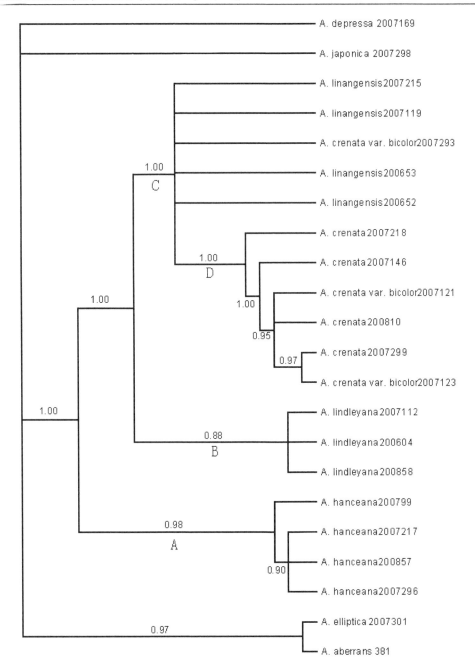

Fig. 2. Phylogenetic tree obtained from Bayesian inference of ITS Sequence data for 22 taxa of the *A. crenata* complex (Number above branches represent the values of posterior probability values)

Photo credit: (a)-(d) Jun Wang; (e) Zhong Wang

Fig. 3. Images of the members of *Ardisia* complex. (a)*Ardisia crenata* Sims; (b) *A. crenata* var. *bicolor* (E. Walker) C. Y. Wu & C. Chen; (c) *A. linangensis* C. M. Hu; (d) *A. hanceana* Mez; (e) *A. lindleyana* D. Dietr.

4. Discussion

The topology in the analysis of two methods (MP, Bayesian) above was consistent based on molecular data, the differences were the bootstrap support values. These results showed that the *A. crenata* complex could be divided into three major groups, which were strongly supported by two phylogenetic methods (MP, Bayesian). The 4 samples of *A. hanceana* were clustered into Clade A and The Clade B was composed of 3 samples of *A. lindleyana*, each one became a true clade in the ITS phylogenetic trees, Clade A was at the base of the trees and the Clade B following closely, which suggested that *A. hanceana* and *A. lindleyana* were divergent earlier than the other three taxa of this complex. From the morphological analysis, *A. hanceana* is different from *A. crenata* in the compound corymbose cyme (vs. umbellate or cymose) inflorescences, 5-16 pieces leaves of flowering branches (vs. 0-5 pieces), sepal almost without punctuate (vs. yellow-brown), fruit diameter 7-9 mm (vs. 5-6mm). *A. lindleyana* also can be distinguished from *A. crenata* with the umbel inflorescences (vs. umbellate or cymose), entire or dentate leaf margin (vs. crisped or crenate), obtuse lobe apex (vs. acute), obtuse triangular ovate sepal (vs. triangular ovate), with sepal and pedicel indumentums (vs. without), so we could separate *A. hanceana* and *A. lindleyana* from the other three taxa of this complex. 4 samples of *A. crenata*, 3 samples of *A. crenata* var. *bicolor* and 4 samples of *A. linangensis* clustered into Clade C. The phylogenic analysis based on ITS with two methods indicated that these three taxa differentiated later than *A. hanceana* and *A. lindleyana*, they might be the same ancestor yet subsequently divergent in different evolutionary patterns almost at the same time. Although all samples of *A. linangensis* not clustered into one Clade, we should pay attention to the Clade D, which all samples of *A. crenata* fell into and it was well supported with (MPBPs/BPP: 95/1.00). This indicated that there has difference between *A. crenata* and *A. linangensis* from the molecular data. From morphological research, we could know there are many characters different between them, *A. linangensis* differ from *A. crenata* in leaf margin (entire or dentate vs. crisped or crenate), marginal punctuate (uniserial vs. irregular), leaf margin nodules (without or on teeth, sparsely vs. between teeth, densely), inflorescences (corymbose or umbellate vs. umbellate or cymose), lobe apex (obtuse vs. acute) and its distribution area is so narrow, can only be seen in south area of Nanling Mountain, which is the nomenclature origin of the species. 2 samples of *A. crenata* var. *bicolor* clustered in Clade D with all samples of *A. crenata*, this indicated that they have very close phylogenetic relationship. The morphological characters of *A. linangensis* and *A. crenata* var. *bicolor* are also very close, they are composed of a multiple branches in the two phylogenetic tree, but *A. linangensis* differ from *A. crenata* var. *bicolor* in many characters such as inflorescences, leaf margin nodules, leaf margin, fruit diameter (Tab. 2) and they are also have different distribution. As for *A. crenata* var. *bicolor*, it was very close to *A. crenata* except for the colour of the leaf. In conclusion, we agree with the opinion of allocating *A. crenata* var. *bicolor* in *A. crenata* and do not support the idea of making *A. linangensis* as the synonym of *A. crenata*. If possible, we want to further research on the population of *A. crenata* complex and find more information used for taxonomy study.

5. References

Angiosperm Phylogeny Group. 2009. An update of the Angiosperm Phylogeny Group classification for the orders and families of flowering plants: APG III. *Botanical Journal of Linnean Society*, 161: 105-121.

Akaike H. 1974. A new look at statistical model identification. *IEEE Transactions on Automatic Control*, 19: 716-723.

Chen C. 1979. Myrsinaceae. Flora Reipublicae Popularis Sinicae Tomus 58. Beijing: Science Press, 1-147. (in Chinese)

Chen C, Pipoly J. J. 1996. III. Myrsinaceae. In Wu Z Y, Raven P H. Flora of China vol. 15: Science Press, 1-38.

Conover C.A. and Poole R.T. 1989. Production and use of *Ardisia crenata* as a potted foliage plant. *Foliage Digest*, 12 (4): 1-3.

Doyle J. J. and Doyle J. L. 1987. A rapid DNA isolation procedure for small quantities of fresh leaf material. *Phytochemical Bulletin*, 19: 11-15.

Felsenstein J. 1985. Confidence limits on phylogenies: an approach using the bootstrap. *Evolution*, 39: 783-791.

Hao G. et al. 2004. Molecular phylogeny of *Lysimachia* (Myrsinaceae) based on choloplast *trn*L-F and nuclear ribosomal ITS sequences. *Molecular Phylogenetics and Evolution*, 31: 323-339.

Hu C. M. 1992. New and noteworthy species of Myrsinaceae from China and Vietnam. *Botanical Journal of South China*, I: 1-13.

Huelsenbeck J. P. and Ronquist F. 2001. MrBayes: Bayesian inference of phylogeny. *Bioinformatics*, 17: 754-755.

Kobayashi H. and Mejia E. 2005. The genus *Ardisia*: A novel source of health-promoting compounds and phytopharmaceuticals. *Journal of Ethnopharmacology*, 96: 347-354.

Lee A. K. 1998. Ecological studies on *Ardisia* native to Korea and the significance as a potential indoor horticultural crop. Ph.D. Dissertation. Dankook University, Cheonan, Korea.

Posada D. and Crandall K. A. 1998. Modeltest: testing the model of DNA substitution. *Bioinformatics*, 14 (9): 817-818.

Posada D. and Buckley T. R. 2004. Model selection and model averaging in phylogenetics: advantages of the AIC and Bayesian approaches over likelihood ratio tests. *Systematic Biology*, 53: 793-808.

Ronquist F. et al. 2003. Bayesian phylogenetic inference under mixed models. *Bioinformatics*, 19: 1572-1574.

Swofford. D. L. 2001. PAUP*: Phylogenetic Analysis Using Parsimony and other methods 4.0b10. Sinauer Associates, Sunderland, M.A.

Thomson J. D. et al. 1997. The Clustal X windows interface: flexible strategies for multiple sequence alignment aided by quality analysis tools. *Nucleic Acids Research*, 25: 4876-4882.

Walker E. H. 1940. A revision of the eastern Asiatic Myrsinaceae. *The Philippine Journal of Science*, 73(1-2): 1-258.

White T. J. et al. 1990. Amplification and direct sequencing of fungal RNA genes for phylogenetics. In "PCR protocols: a guide to methods and applications" (Innis M., Gelfand D., Sinisky J. and White T. Eds.), Acadmic Press: San Diego, CA, 315-322.

Yunnan Institute of Botany. 1977. Flora of Yunnanica Tomus 1. Beijing: Science Press, 332-361. (In Chinese)

9

Plant Special Cell – Cotton Fiber[1]

Ling Fan, Wen-Ran Hu, Yang Yang and Bo Li

Institute of Nuclear and Biological Technologies,
Xinjiang Academy of Agricultural Sciences, Nanchang Road, Urumqi
China

1. Introduction

Cotton fibers are single-celled outgrowths from individual epidermal cells on the outer integument of the ovules in the developing cotton fruit. Fibers of upland cotton (*G. hirsutum* L.) generally grow up to 30 to 40 mm in length and ~15 μm in thickness at full maturity. Their development consists of four overlapping stages: fiber initiation, cell elongation, secondary wall deposition, and maturation. The thickened secondary walls of mature cotton fibers have long been considered unique in that they were thought to consist of nearly pure cellulose and to be devoid of hemicellulose and phenolics. However, other plant derived fibers such as flax (*Linum usitatissimum* L) and ramie (*Boehmeria nivea* L) fibers have been shown to contain phenolics except cellulose (Angelini et al., 2000; Day et al., 2005).

Plant cell wall phenolics consist of two groups of compounds: (1) lignin, the polymer of monolignol units, linked by oxidative coupling; and (2) low molecular weight hydroxycinnamic acids, that are bound to various cell wall components and are involved in cross-linkages (Iiyama at al., 1994; Wallace & Fry 1994). From a functional point of view, plant cell wall phenolics protect cellulose fibers in plant cell wall from chemical and biological degradation (Grabber et al., 2004) and can influence wall mechanical strength, growth, morphogenesis and responses to biotic and abiotic stresses (Wallace & Fry, 1994; Boerjan et al., 2003).

The emergence of lignin during evolution is believed to be a crucial adaptation for terrestrial plants from aquatic ancestors(Kendrick & Crane, 1997; Peter & Neale, 2004). It is mainly present in secondary thickened cell walls within xylem tissues, where it provides rigidity and impermeability to the cell walls.Later on, "lignin-like"compounds have been identified in primitive green algae (Delwiche et al., 1989), and lignins have been found exist within a red alga's calcified cells that lack hydraulic vasculature and have little need for additional support (Martone et al., 2009). Lignin is also an integral constituent of the primary cell walls of the dark-grown maize (Zea mays L.) coleoptile(Müsel et al., 1997), elongation zone of maize primary root (Fan et al., 2006), the juvenile organs those are still in the developmental state of rapid cell extension. Molecular and biochemical evidences have also showed phenylpropanoid synthesis and presence of wall-linked phenolics in white, soft cotton

[1] Project supported by the Joint Funds of the National Natural Science Foundation of China (U1178305) , Hi-Tech Research and Development Program of Xinjiang, China (201111116), National Natural Science Foundation of China (31060173).

fibers. The aim of this chapter is to present our new results about cotton fiber growth and development. Molecular, spectroscopic and chemical techniques were used to prove the possible occurrence of previously overlooked accumulation of phenolics during secondary cell wall formation in cotton fibers.

2. Cotton fiber growth

2.1 Cotton fiber elongation

North Xinjiang in China is an upland cotton (*G. hirsutum* L.) growing area. Local effective accumulated temperature (≥10°C) is less than 3500°C. Frost-free period is less than 150 d. Under this growing condition, we observed the fiber cell elongation of the cultivar Xinluzao36 (*G. hirsutum* L.) and cultivar Xinhai 22 (*G. barbadense* L). The fiber length was determined kinetically by the method of Gokani and Thaker (2000). The final values were taken as means (n=6 balls×20 seed with fibers). The fiber elongation of both varieties went through slow-fast-slow-stop process and showed typical sigmoid growth curve (Fig. 1). During 0-10 DPA, fibers were initiating and elongating very slowly in 0-5 DPA, and fibers were accelerately elongating in 5-10 DPA. During 10-15 DPA, the fiber elongation rate was actively accelerated. Both varieties showed similar fiber elongation rate in 0-15 DPA. However, the fiber elongation rate of Xinhai 22 was continuously actively accelerated until 19 DPA. The fiber elongation rate was decelerated during 15-21 DPA for Xinluzao36 and during 19-23 DPA for Xinhai 22 respectively. The results suggested that longer-fiber-length variety Xinhai 22 not only had longer fiber elongation period, but also had longer actively accelerated fiber elongation period.

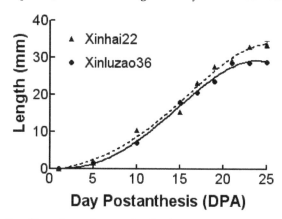

Fig. 1. Kinetics of cotton fiber elongation postanthesis

2.2 Cotton fiber secondary wall deposition

The Ultra-structural developmental process of fiber cell from early maturing cotton cultivar Xinluzao 36 was observed by using transmission electron microscopy. At 10 DPA, when the primary cell wall of cotton fiber was thin and even thickness, and a big vacuole located at the central of fiber cell, there were rich organelles, such as mitochondria, ribosomes and Golgi bodies in the cytoplasm (Fig. 2 A-D). At 20 DPA, a thin layer of the secondary cell wall formed inside the primary cell wall clearly and a part of the organelles disappeared (Fig. 2 E-F). Subsequently, the secondary cell wall thickened rapidly (Fig. 3). The average thickness increased around 0.14 μm per day from 30 DPA to 40 DPA and around 0.47 μm per day from 40 DPA to 50 DPA (Fig. 4). The secondary cell wall gradually thickened and formed daily growth rings (Fig 3. E). Then the vacuum inside fiber cell became a narrow gap and the thickness of cell wall became thinner as the fiber maturing and water loss (Fig 3. F). Results showed that the fiber development process of early-maturing cotton was quite similar to other varieties reported (Xu, 1988), despite flowering and maturing earlier.

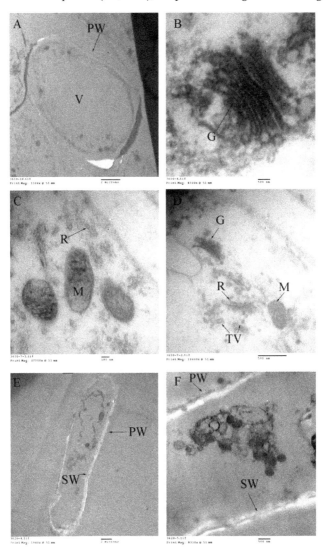

Fig. 2. Transverse section characters of 10 DPA and 20DPA fiber cell
10DPA:A(×6000),B(×15000),C(×60000); 20DPA:E(×3500),F(×15000); DPA(Day post anthesis); PW(Primary wall); G(Golgi apparatus); M(Mitochondrion); R(Rough endoplasmic reticulum); SW(Secondary wall); TV(Transport vesicles); V(Vacuole)

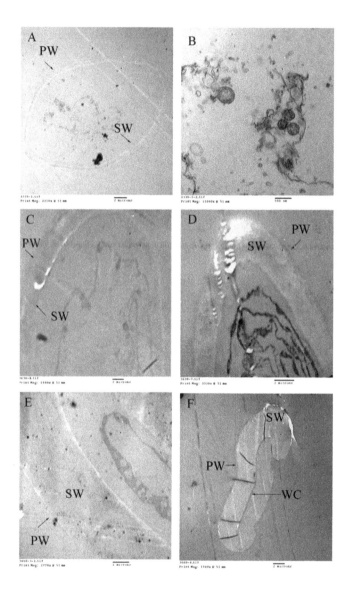

Fig. 3. Transverse section characters of 30 DPA ,40 DPA, 50 DPA and mature fiber cell
30DPA:A(×4000),B(×20000); 40DPA:C(×3500); 50DPA:E(×5000); Mature: F(×15000);DPA(Day
post anthesis); PW(Primary wall); SW(Secondary wall); MC(Mesocoele).

2.3 Ovule culture fed by middle product in phenylpropanoid pathway

Based on the system of ovule cultured in vitro, ferulic acid (FA), which is middle product in phenylpropanoid pathway, was fed in the media for either fertilized or unfertilized ovules. With the assistance of image digitization technology (Fan et al. 2011), the growth condition of the fibers at different period was analyzed. The results revealed that higher concentration of exogenous FA (100 µmol/L) inhibited the normal growth of the fibers(not show), while low level of FA (50 µmol/L) had not showed such inhibited phenomenon, in the opposite, fiber elongation was accelerated (Fig. 5). The results showed that FA had the effects on the cotton fiber growth by adjusting the FA concentration in the media. The turning point for accelerating and decelerating growth of cotton fiber was possibly at the concentration ranged between 50 µmol/L and 100 µmol/L. TEM observation on fiber transverse section indicated that the cell wall thickness was significantly increased by exogenous FA(P<0.01). In addition, acetyl bromide chemical analysis revealed that the content of phenylpropanoid compound in cotton fiber was increased 5.2% for the fibers fed with 50 µmol/L FA. The investigation of cotton fiber growth, cell wall thickness and chemical analysis revealed that exogenous FA had significant effects on cotton fiber development in ovule culture system. This work improved our understanding of cotton fiber development mechanism and the effect of phenylpropanoid pathway and its products on cotton fiber development.

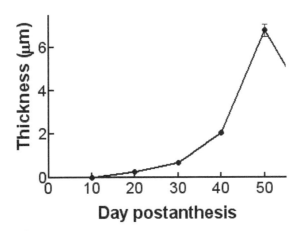

Fig. 4. Increasing thickness of secondary wall in fiber development
15 images of cell wall from each period were randomly selected, the perpendicular thickness of the secondary wall of 15 evenly distributed points at each image were measured.Values =Means ± SE, n = 225 measurements.

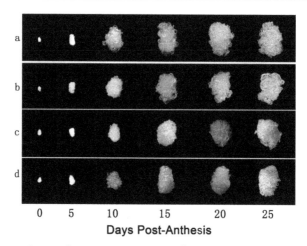

Days Post-Anthesis

Fig. 5. The representative ovule expansion images of FA treatment under time course (a) represents fertilized ovules; FA final concentrations of 50μM; (b) represents fertilized ovules control (no FA adding); (c) represents unfertilized ovules; FA final concentrations of 50μM; (d) represents unfertilized ovules control.

3. Cotton fiber secondary wall development

3.1 Genes in phenylpropanoid pathway were expressed in developing cotton fibers

To explore the expression regularity of genes in phenylpropanoid pathway in developing cotton fibers and to investigate the relationships between cotton fiber quality and genes expression at transcriptional level, the relative quantitative expression of genes in phenylpropanoid pathway, such as *GhPAL*, *Gh4CL*, *GhC4H*, *GhCOMT*, *GhCCoAOMT*, *GhCCR* and *GhCAD*, were studied by semi-quantitative reverse transcription-polymerase chain reaction (RT-PCR) in different organs and different time of developmental cotton fibers respectively (Fig. 6). The expression of *GhC4H*, *GhCOMT and GhCCoAOMT* had higher levels in both root and stem. Along with the initiation of fiber cell wall thickening, most of the genes displayed high expression levels. The expression of *GhCAD1*, *GhCCR2* and *GhCCR3* were gradually reduced in the development of cotton fiber, but they were not have significant differences in different quality of cotton fiber in all developmental stages.

Since CAD has been considered a key enzyme in the phenylpropanoid biosynthesis pathway(Steeves et al. 2001, Boerjan et al. 2003), *GhCADs* expressions in cotton fiber were analyzed. Relative quantitative RT-PCR analysis was carried out at the time of secondary cell wall development (25 DPA) in the cotton fiber cell using the gene-specific primers. The analysis showed that *GhCAD6* and *GhCAD1* were predominantly expressed among seven gene homologs. The relative expression of *GhCAD6* was increased during the period of secondary wall development in cotton fibers, and reached the higher level at 20 DPA.

The relationship between phenylpropanoid compounds and cotton fiber development was explored in our laboratory. The higher expression time of key enzyme in phenylpropanoid pathway *CAD* genes was coincident with the secondary wall development of both the fiber cells and the organs with vascular elements. In addition, the time of *GhCAD6* higher expression coincided with the time of secondary wall formation of cotton fiber (Fig. 7). This is a kind of indicator for the biosynthesis of the phenylpropanoid unit and cell wall phenolics. Those results explained that the pathway was active not only in vascular tissues but also in fiber cells with secondary wall thickening. Guo et al. (2007) found that the regulation of seven metabolic pathways including pathways of phenol metabolism, showed significant changes during the maturation of cotton fiber. Recent research by Hovav et al. (2008) found that genes of the phenylpropanoid metabolism in cluster 6 gradually increase in expression during fiber development. Their studies provide indirect support for our findings.

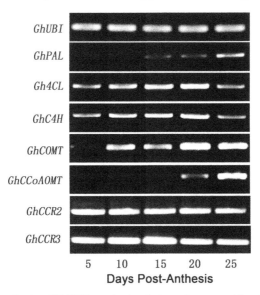

Fig. 6. Relative quantitative RT-PCR analysis of phenylpropanoid pathway genes in developing cotton fiber

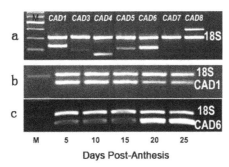

Fig. 7. Relative quantitative RT-PCR analyses of *GhCADs* in developing cotton fiber. (a) Relative quantitative RT-PCR analyses of seven *GhCAD* gene homologs at 25 DPA of developing cotton fiber; (b) Relative quantitative RT-PCR analyses of *GhCAD6* as time course; (c) Relative quantitative RT-PCR analyses of *GhCAD1* as time course. Lane M shows nucleotide markers; 18S indicates a 315 bp fragment of ribosomal RNA used as the internal standard.

3.2 Cloning and characterization of genes in phenylpropanoid pathway in developing cotton fibers

Sixteen gene complete cDNAs in this pathway had been cloned from developing cotton fibers. Their amino acid sequences had high homology with those corresponding genes from other plants (Table 1).

Genes	NCBI No.	Genes	NCBI No.
GhCAD1	EU281304	Gh4CL1	FJ479707
GhCAD6	EU281305	GhCOMT1	FJ479708
GhCAD3	FJ376601	GhCOMT2	FJ479709
GhCAD7	FJ376602	GhC4H1	FJ848866
GhCCR1	FJ376603	GhC4H2	FJ848867
GhCCR4	FJ376604	GhCAD5	FJ848868
GhCCoAOMT1	FJ376605	GhCOMT3	FJ848869
GhCCoAOMT2	FJ376606	Gh4CL2	FJ848870

Table 1. Genes cloned from developing cotton fiber and their NCBI numbers

3.2.1 Gene of initiation (*GhPAL*)

PAL is the first enzyme of the general phenylpropanoid pathway and catalyzes the nonoxidative deamination of phenylalanine to trans-cinnamic acid and NH3. Subsequent enzymatic steps involving the actions of cinnamate 4-hydroxylase (C4H), 4-coumarate:CoA ligase (4CL), hydroxycinnamoyl-CoA transferase (HCT), p-coumarate 3-hydroxylase (C3H), caffeoyl-CoA O-methyltransferase (CCoAOMT), cinnamoyl-CoA reductase (CCR), ferulate 5-hydroxylase (F5H), caffeic acid O-methyltransferase (COMT), and cinnamyl alcohol dehydrogenase (CAD) catalyze the biosynthesis of monolignols (Andersen et al.2008). Semi-quantitative RT-PCR analysis revealed that *GhPAL* gene was differentially expressed in different developmental stages, Along with the initiation of fiber cell wall thickening, *GhPAL* gene displayed high expression levels (Fig. 8).

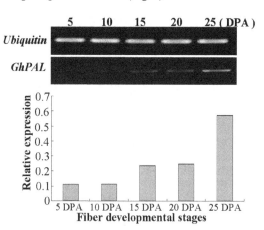

Fig. 8. Expression patterns of *GhPAL* in developing cotton fibers

3.2.2 Genes of methylation reaction (*GhCOMT* and *GhCCoAOMT*)

Caffeic acid O-methyltransferase (COMT) and caffeoyl-CoA-3-O-methyltransferase (CCoAOMT) genes encode two methyltransferases at substrate levels in lignin biosynthesis (Yoshihara et al.2008). COMT is essential for syringyl lignin (S unit) biosynthesis and has long been considered as the only methylating enzyme involved in lignification; CCoAOMT is involved in the biosynthesis of guaiacyl (G) and syringyl (S) lignins.

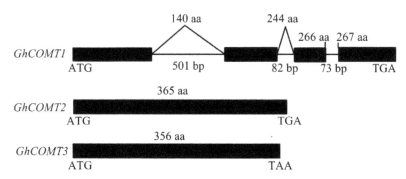

Fig. 9. The structures of *GhCOMT1*, *GhCOMT2* and *GhCOMT3* genes in cotton

Exons are denoted by black boxes. Introns are denoted by lines. The lengths of the introns in base pairs are indicated. The number at the boundaries of each exon indicates the codon at which the intron is located. The translation initiation and termination codons are shown. aa, amino acids.

Full-length cDNA of a key enzyme genes *GhCOMT1*, *GhCOMT2* and *GhCOMT3* related to lignin metabolism in cotton (*G. hirsuturm L.*) were isolated, and their cDNA and the amino acid sequences were analyzed by bioinformatics methods (Fig. 9). Semi-quantitative RT-PCR analysis revealed that *GhCOMT1*, *GhCOMT2* and *GhCOMT3* were differentially expressed in different tissues, and *GhCOMT1* and *GhCOMT2* mRNA accumulated most abundantly in root, *GhCOMT3* was highly expressed in stem. For the fiber developing, from 5 to 25 DPA, *GhCOMT1* and *GhCOMT3* were consistently expressed, while *GhCOMT2* was increasingly expressed.

Two *GhCCoAOMT* genes were cloned in developing cotton fiber. Semi-quantitative RT-PCR analysis revealed that *GhCCoA0MT1* and *GhCCoA0MT2* can be expressed in different kinds of cotton tissues (Fig. 10), and their mRNA accumulated most abundantly in stem *GhCCoAOMT1* expression in cotton tissues was stem> root> petal > hypocotyl>10 DPA fiber> stamen> ovule> leaf. Meanwhile, the expression peak of *GhCCoAOMT1* appeared at 25 DPA, while *GhCCoAOMT2* was at 10 DPA and 15 DPA (Fig. 11). The results confirmed that all of the genes of the two methyltransferase gene families had higher expression quantity, which was coincident with vascular tissues, such as root and stem, while tissue specific and development period specific expression patterns were detected in other orgens and growing fibers.

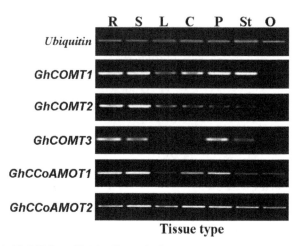

Tissue type

Fig. 10. *GhCOMT1, GhCOMT2, GhCOMT3* and ubiquitin gene RT-PCR products from different cotton tissues. R: Root; S: Stem; L: Leaf; H: Hypocotyl; P: Petal; St: Stamen; O: Ovule.

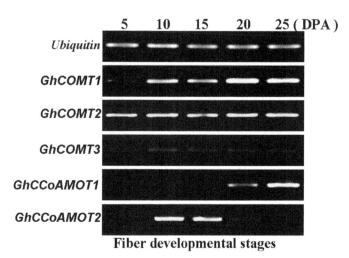

Fiber developmental stages

Fig. 11. *GhCOMT1, GhCOMT2, GhCOMT3* and ubiquitin gene RT-PCR products from different developmental stages

3.2.3 Genes of reduction reaction(*GhCCR* and *GhCAD*)

Cinnamoyl-CoA reductase (CCR), one of the key enzymes in the first step of the phenylpropanoid pathway, catalyzes the conversion of cinnamoyl-CoA esters to their respective cinnamaldehydes and is the first enzyme of the monolignolspecific part of the lignin biosynthetic pathway; Cinnamyl alcohol dehydrogenase(CAD) is the last enzyme on the pathway to the monolignols coniferyl and sinapyl alcohols, from which lignins are normally derived (Goujon et al., 2003).

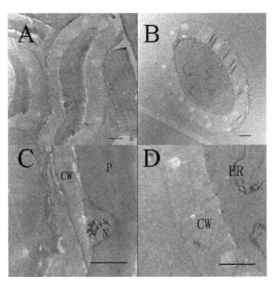

Fig. 12. Effects of the GhCCR4 on secondary wall thickness of fibers under a transmission electron microscope. A:Check; B:Transgenic; C:Magnif of check; D: Magnif of transgenic; N: Nucleus; P: Pusule; CW: Cell Wall; ER: Endocytoplasmic Reticulum; Bar=2μm.

In order to analyze the function of these genes, five GhCAD and two GhCCR genes were isolated from cotton fiber, and transient expression vector of cotton GhCCR4 gene was constructed. The transient expression vector pGUS-GhCCR4 are driven by 35s promoter with GUS reporter gene and the target gene, GUS reporter gene and target gene simultaneously express in single cell. They were transformed into cotton ovule using PDS-1000/He biolistic particle delivery system. The results indicated that the transient expression vector pGUS-GhCCR4 could be high efficiency expressed in the epidermal cells of cotton fiber. Transmission electron microscopy demonstrated that the wall thickness of transgenic fiber was increased to 17% of that of the wild type (Fig. 12). These findings suggest that GhCCR4 could play a critical role in the processes of secondary cell wall formation during fiber development.

4. Cotton fiber chemical component

The cotton fiber wall phenolics were observed by UV induced auto-fluorescence (Leica DMI6000B microscope) with co-observation by scanning electron microscopy (SEM). The washed cotton fiber, especially the cut ends, clearly showed the green-blue auto-fluorescence indicative of wall phenolics (Fig. 13 A). The cotton fiber residues after extraction of lignin-like phenolics by the thioglycolate method showed less green-blue auto-fluorescence (Fig. 13 B). Apparently, the thioglycolate method removed some but not all of the phenolic compounds from the cell walls of the cotton fibers. Notably, the autofluorescence of the remaining residues became lower after saponified phenolics were extracted (Fig. 13 C). Moreover, the secondary layers of cell wall separated and the residual fibers lost their tubular shape (Fig. 13 F).

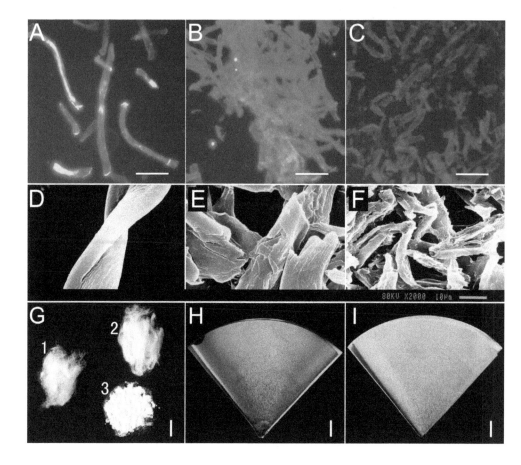

Fig. 13. Microscopic observation of cotton fibers and their residues.
(A) Auto-fluorescence of cut cotton fiber; (B) Auto-fluorescence of the residues left after thioglycollate extraction; (C) Auto-fluorescence of the fiber residues after further extraction of saponifiable phenolics; (D) Scanning electron microscopy (SEM) observation of cotton fiber; (E) SEM observation of the residues left after thioglycollate extraction; (F) SEM observation of the fiber residues after further extraction of saponifiable phenolics; (G) Cotton fiber before washing (1), after washing (2), and residues after thioglycollate extraction (3); (H) The brown insoluble residues left on the filter paper by Klason extracts of cotton fiber; (I) Decreased amount of brown insoluble residue after thioglycolate extraction; Horizontal bars in color figures represent 100 µm, and in black and white figures represent 10 µm. Vertical bars represent 1 cm.

The classical methods of analyzing lignin content including Klason, thioglycolate and acetyl bromide were optimized in order to suit for analyzing cotton fiber. We analyzed the contents of lignin-like phenolics in mature cotton fiber samples using the optimized Klason, thioglycolate, and acetyl bromide methods (Table 2). Although the contents varied between samples and test methods, all of the analyzed samples (14 cultivars and two lines) of cotton fiber contained lignin-like phenolics (Table 2). With the Klason method, 72% H_2SO_4 dissolved away the polysaccharides and acid soluble small molecular phenolics, leaving lignin-like phenolics as an insoluble residue. The brown insoluble residue from cotton fiber clearly showed on the filter paper (Fig. 13 H). The content of Klason phenolics in cotton fiber was relative lower (0.37–1.08%). The thioglycolate method only extracted partial phenolics from cotton fiber (Fig. 13 B,I). There was still some brown residue left after thioglycolate extraction (Fig. 13 I). Therefore, the content of thioglycolate phenolics in cotton fiber was quite low (0.13–0.35%). The acetyl bromide method might reveal both lignin and hydroxycinnamic acids in cotton fibers, since the cotton fibers were completely digested. The content of acetyl bromide phenolics was 2.23–2.63%. The content of lignin-like phenolics in cotton fiber in single boll was kinetically increased with the fiber developing during 20DPA to mature, which had the same trend as the secondary wall thickening by ultra-structure observation(Fig. 14). Compared with lignin accumulation and secondary wall thickness observed by ultra-structure, we found the same increasing trend. The results suggested that monolignol biosynthesis and wall-linked lignin-like phenolics involved in the secondary wall thickening of cotton fibers (Fan et al., 2009).

No.	Sample	Origin	Planting place	Lignin-like phenolics in cotton fiber(%)		
				Thioglycolate	Acetyl bromide	Klason
1	Xinluzao6	China	Manasi,China	0.330±0.006	2.558±0.018	0.902±0.044
2	CJ01	China	Manasi,China	0.252±0.004	2.353±0.004	0.683±0.068
3	Liao3206	China	Manasi,China	0.249±0.009	2.420±0.037	0.468±0.025
4	Tashkent3	Uzekistan	Manasi,China	0.217±0.016	2.578±0.033	0.713±0.013
5	Zhong35	China	Kuche,China	0.172±0.007	2.455±0.032	0.663±0.003
6	Zhong49	China	Kuche,China	0.127±0.010	2.566±0.040	0.675±0.005
7	KK1543	USSR	Kuche,China	0.230±0.008	2.472±0.015	0.783±0.045
8	MC50	USA	Kuche,China	0.347±0.017	2.602±0.017	0.660±0.010
9	CB1135	USA	Kuche,China	0.184±0.008	2.234±0.043	0.703±0.015
10	DP2156	USA	Kuche,China	0.281±0.023	2.633±0.021	0.723±0.038
11	NC33B	USA	Anyang,China	0.340±0.034	2.465±0.025	0.578±0.018
12	NC32B	USA	Anyang,China	0.207±0.012	2.545±0.030	0.738±0.013
13	NC20B	USA	Anyang,China	0.146±0.024	2.578±0.020	0.502±0.019
14	DP99B	USA	Anyang,China	0.190±0.017	2.449±0.030	0.860±0.030
15	DPH37B	Hybrid	Anyang,China	0.197±0.023	2.336±0.020	0.370±0.007
16	Xuzhou142	China	PKU,China	0.169±0.021	2.601±0.058	1.078±0.033

Table 2. The content of lignin-like phenolics in cotton fiber tested by Thioglycolate, Acetyl bromide and Klason methods

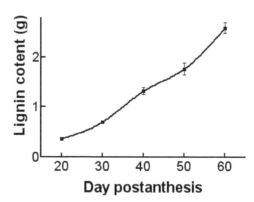

Fig. 14. Kinetics of lignin content of cotton fiber in single boll.

The samples were taken from different fiber developmental stages. The fiber were carefully detached from the seeds. The fiber samples were washed twice with homogenization buffer (50 mmol/L Tris-HCl, 10 g/L Triton X-100, 1 mol/L NaCl; pH8.3), twice with 80% (v/v) acetone and once with acetone as described Niklas(2000). The lignin content of 2.00g of cotton fiber was analysed by Klason method. Values =Means ± SE, n = 3 measurements.

5. Conclusion

The mature cotton (G. hirsutum L.) fiber have long been considered unique in that their thickened secondary cell were thought to consist of nearly pure cellulose. However, we found that secondary wall deposition in cotton fiber and in water-conducting xylem cells shares common elements. RT-PCR analysis showed that genes in the monolignol biosynthesis pathway were expressed in the secondary wall formation of cotton fiber. Sixteen gene complete cDNAs in this pathway had been cloned from developing cotton fibers. Their amino acid sequences had high homology with those corresponding genes from other plants. Some of these genes were predominantly expressed during secondary wall formation in cotton fibers. Chemical analysis confirmed the presence of lignin-like phenolics in mature cotton fiber from germplasm resources in different areas of China, USA, Australia, Russia and Mexico. We concluded that monolignol biosynthesis and wall-linked lignin-like phenolics involved in the secondary wall thickening of cotton fibers.

6. References

Andersen, J.R.; Zein, I.; Wenzel, G.; Darnhofer, B.; Eder, J.; Ouzunova, M. & Lübberstedt T. (2008). Characterization of phenylpropanoid pathway genes within European maize (Zea mays L.) inbreds, *BMC Plant Biology*, Vol.8, No.2, pp.1-14

Angelini, L.G.; Lazzeri, A.; Levita, G.; Fontanelli, D. & Bozzi, C. (2000). Ramie (*Boehmeria nivea* (L.) Gaud.) and Spanish Broom (*Spartium junceum* L.) fibres for composite materials: agronomical aspects, morphology and mechanical properties. *Ind Crop Prod*, Vol.11,(March 2000), pp. 145–161

Boerjan, W.; Ralph, J. & Baucher, M. (2003). Lignin biosynthesis. *Annu Rev Plant Biol*, 54: pp.519-546

Day, A.; Ruel, K.; Neutelings, G.; Crônier, D.; David, H.;, Hawkins, S. & Chabbert, B. (2005). Lignification in the flax stem: evidence for an unusual lignin in bast fibers. *Planta,* Vol.222, No.2, (Jun 21), pp. 234–245

Delwiche, C.F.; Graham, L.E. & Thomson, N. (1989). Lignin-like compounds and sporopollenin in *Coleochaete*, an algal model for land plant ancestry. *Science,* Vol.245, No.4916, (July), pp. 399–401

Fan, L.; Linker, R.; Gepstein, S.; Tanimoto, E.; Yamamoto, R. & Neumann, P.M. (2006). Progressive Inhibition by Water Deficit of Cell Wall Extensibility and Growth Along the Elongation Zone of Maize Roots Is Related to Increased Lignin Metabolism and Progressive Stelar Accumulation of Wall Phenolics. *Plant Physiol,* Vol.140, No.2, pp. 603-612

Fan, L.; Shi, W.J.; Hu, W.R.; Hao, X.Y.; Wang, D.M.; Yuan, H. & Yan, H.Y. (2009). Molecular and biochemical evidence for pheylpropanoid synthesis and presence of wall-linked phenolics in cotton fiber. *J Integr Plant Biol,* Vol.51, No.7, pp. 626-637

Fan, L.; Lü, M.; Ni, Z.Y.; Hu, W.R. & Wang, J. (2011). Digital image analysis of expansion growth of cultured cotton ovules with fibers and their responses to ABA. *In Vitro Cell Dev Biol-Plant,* Vol.47, No.3, pp. 369-374

Gokani, S.J. & Thaker, V.S. (2000). Physiological and biochemical changes associated with cotton fiber development. VIII. Wall components. *Acta Physiol Plant,* Vol.22, No.4, pp.403-408

Goujon, T.; Sibout, R.; Eudes, A.; MacKay, J. & Jouanin, L. (2003). Genes involved in the biosynthesis of lignin precursors in Arabidopsis thaliana. *Plant Physiology and Biochemistry,* Vol.41, No.8, pp.677–687

Grabber, J.H.; Ralph, J.; Lapierre, C. & Barrière, Y. (2004). Genetic and molecular basis of grass cell-wall degradability. I. Lignin–cell wall matrix interactions. *C. R. Biol,* Vol. 327, No.5, (May), pp.455–465

Guo, J.Y.; Wang, L.J.; Chen, S.P.; Hu, W.L. & Chen, X.Y. (2007). Gene expression and metabolite profiles of cotton fiber during cell elongation and secondary cell wall synthesis. *Cell Res,* Vol. 17, No.5, (May), pp. 422-434

Hovav, R.; Udall, J.A.; Hovav, E.; Rapp, R.; Flagel, L. & Wendel, J.F. (2008). A majority of cotton genes are expressed in single-celled fiber. *Planta,* Vol. 227, Vol. 2, pp. 319-329.

Iiyama, K.; Lam, T.B.T. & Stone, B.A. (1994). Covalent cross-links in the cell wall. *Plant Physiol,* Vol.104, No.2, (February), pp. 315 – 320

Kendrick, P. & Crane, P.R. (1997). The origin and early evolution of plants on land. *Nature,* Vol.389, Vol.6646,(April), pp. 33–39

Martone, P.T.; Estevez, J.M.; Lu, F.; Ruel, K.; Denny, M.W.; Somerville, C. & Ralph, J. (2009). Discovery of lignin in seaweed reveals convergent evolution of cell-wall architecture. *Cur Bio,* Vol.19, No.2, (January), pp.169–175

Müsel, G.; Schindler, T.; Bergfeld, R.; Ruel, K.; Jacquet, G.; Lapierre, C.; Speth, V. & Schopfer, P. (1997). Structure and distribution of lignin in primary and secondary cell walls of maize coleoptiles analyzed by chemical and immunological probes. *Planta,* Vol.201, No.2, pp.146–159

Niklas, K.J.; Freaner, F.M.; Ojanguren, C.T. & Paolillo, D.J. (2000). Wood biomechanics and anatomy of Pachycereus Pringlei. *American Journal of Botany,* Vol.87, No.4, (April), pp.469-481

Peter, G. & Neale, D. (2004). Molecular basis for the evolution of xylem lignification. *Cur Opin Plant Biol* , Vol.7, No.6, (December), pp.737–742

Steeves, C.; Förster, H.; Pommer, U. & Savidge, R. (2001). Coniferyl alcohol metabolism in conifers-I. Glucosidic turnover of cinnamyl aldehydes by UDPG: coniferyl alcohol glucosyltransferase from pine cambium. *Phytochemistry,* Vol.57, No.7, (August), pp. 1085-1093

Wallace, G. & Fry, S. (1994). Phenolic components of the plant cell wall. *Int Rev Cytol,* Vol.151, pp. 229–267

Xu, C.N.; Yu, B.S.; Zhang, Y.; Jia, J.Z. & Shou, Y. (1988). The comparative studied on fiber development of four cultispecies in cotton. *Acta Agriculturae Universitatis Pekinensis,* Vol.14, pp. 113-119

Yoshihara, N.; Fukuchi-Mizutani, M.; Okuhara, H.; Tanaka, Y. & Yabuya, T. (2008). Molecular cloning and characterization of O-methyltransferases from the flower buds of Iris hollandica. *J Plant Physiol,* Vol.165, No.4, (March), pp. 415-422

10

Hybrid Lethality in the Genus *Nicotiana*

Takahiro Tezuka

Graduate School of Life and Environmental Sciences,
Osaka Prefecture University
Japan

1. Introduction

Reproductive isolation is a mechanism that separates species. It is considered to play a crucial role in the evolution of animals and plants. Reproductive isolation is divided into two types of barriers, namely prezygotic and postzygotic. In plants, a typical prezygotic barrier observed after pollination is pollen–pistil incongruity (or incompatibility). Specifically, when the pollen of one species is rejected by the pistil of another species, but the reciprocal cross is successful, the incongruity is called unilateral incongruity. Postzygotic barriers include seed abortion, and hybrid lethality and hybrid sterility in the F_1 generation. When F_1 hybrids are normal but their F_2 progeny contains lethal or sterile individuals, this phenomenon is called hybrid breakdown and is discriminated from abnormalities in the F_1 generation. Whereas these prezygotic and postzygotic barriers contribute to speciation, they are obstacles for plant breeders, especially in breeding programs involving wide hybridization. Hybrid plants from normal parents sometimes show weak growth or die before maturity. Several terms have been used to describe these phenomena, i.e., hybrid lethality, hybrid weakness, hybrid necrosis and hybrid inviability. Hybrid lethality is observed in certain cross combinations in many plant species (Bomblies & Weigel, 2007). In this chapter, I review studies of hybrid lethality in the genus *Nicotiana*.

2. The genus *Nicotiana*

The genus *Nicotiana* (Solanaceae) includes 76 species classified into 13 sections (Knapp et al., 2004). Species in most sections are distributed mainly in the Americas. The exception is section *Suaveolentes*. This section includes 25 species restricted to Australia and islands of the South Pacific, and one African species, *N. africana*, which is the only known species in Africa. These *Suaveolentes* species are geographically isolated from the majority of species in other sections. Many researchers have attempted to reveal the origin and evolution of this complex genus.

Among *Nicotiana* species, only *N. tabacum* ($2n = 48$, SSTT) and *N. rustica* ($2n = 48$, PPUU) are cultivated tobacco species, whereas the others are wild tobacco species. Of these cultivated species, *N. tabacum*, which belongs to section *Nicotiana*, is the most important for commercial purposes. In *N. tabacum* breeding programs, wild species are valuable as sources of disease resistance (Bai et al., 1995; Burk & Heggestad, 1966; Holmes, 1938; Li et al., 2006; Stavely et al., 1973) and cytoplasmic male sterility (Nikova & Zagorska, 1990; Nikova et al., 1991, 1999). Therefore, many interspecific crosses have been conducted between *N. tabacum* and wild species. However, hybrid lethality often presents a barrier to introduction of desirable characteristics into *N. tabacum*.

3. Types of hybrid lethality in *Nicotiana*

Inviable hybrid seedlings in *Nicotiana* initially show specific phenotypes (surface symptoms) depending on cross combinations (Tezuka et al., 2009; Yamada et al., 1999). Hybrid lethality in this genus is classified into five types based on the external phenotypes as follows:

- Type I: browning of shoot apex and root tip
- Type II: browning of hypocotyl and roots
- Type III: yellowing of true leaves
- Type IV: formation of multiple shoots
- Type V: fading of shoot color.

Different physiological processes are considered to be involved in at least Types I–IV hybrid lethality, because whether three methods to rescue inviable hybrids (cultivation at elevated temperatures, cotyledon culture, and cytokinin treatment) are effective or not depends on the lethality type (Yamada et al., 1999) as described later. It is possible different causative factors control different types of hybrid lethality.

4. Hybrid lethality in crosses between *Suaveolentes* species and *N. tabacum*

In *Nicotiana*, hybrid lethality is well studied in crosses between species of section *Suaveolentes* and *N. tabacum*. To my knowledge, 22 species in section *Suaveolentes* have been crossed with *N. tabacum* and the viability of the hybrid seedlings has been reported. The results of the crosses are summarized in Table 1.

Twenty *Suaveolentes* species yield inviable hybrids in crosses with *N. tabacum*. In most cases, the hybrid lethality is of Type II (Iizuka et al., 2010; Laskowska & Berbeć, 2011; Tezuka & Marubashi, 2006b; Tezuka et al., 2006, 2007, 2010). Only hybrid seedlings between *N. occidentalis* and *N. tabacum* show Type V lethality (Tezuka et al., 2009). Types II and V lethality in these crosses is observed at or below 28°C, but is completely suppressed at elevated temperatures ranging from 34 to 36°C.

The remaining two *Suaveolentes* species, *N. benthamiana* and *N. fragrans*, yield 100% viable hybrids in crosses with *N. tabacum* (DeVerna et al., 1987; Iizuka et al., 2010, 2011; Tezuka et al., 2010). These species are exceptions in this section with respect to hybrid lethality.

4.1 Causes of hybrid lethality

Reciprocal hybrids with *N. tabacum* were produced using the above-mentioned 19 *Suaveolentes* species that yield inviable hybrids (excluding *N. wuttkei*). In all these crosses, hybrid lethality was observed regardless of cross direction, which indicates hybrid lethality is a result of the interaction of coexisting heterologous genomes, and not a cytoplasmic effect (Iizuka et al., 2010; Tezuka & Marubashi, 2004, 2006a; Tezuka et al., 2006, 2009, 2010). Subsequent genetic analyses have identified the chromosome and genes responsible for hybrid lethality.

Suaveolentes species	Haploid chromosome number	F$_1$ pheno-type	Suppression at elevated temperatures	Factors responsible for hybrid lethality[a]		References[b]
				In *N. tabacum*	In *Suaveolentes* species	
N. africana	23	Type II lethality	Possible	Q chromosome	ND	7
N. amplexicaulis	18	Type II lethality	Possible	Q chromosome	ND	3
N. benthamiana	19	Viable	–	–	*hla1-2*	1, 6, 8
N. cavicola	23	Type II lethality	Possible	ND	ND	6
N. debneyi	24	Type II lethality	Possible	Q chromosome	*Hla1-1*	4, 8
N. excelsior	19	Type II lethality	Possible	Q chromosome	ND	7
N. exigua	16	Type II lethality	Possible	ND	ND	6
N. fragrans	24	Viable	–	–	*hla1-2*	7
N. goodspeedii	20	Type II lethality	Possible	Q chromosome	ND	7
N. gossei	18	Type II lethality	Possible	Q chromosome	ND	7
N. hesperis	21	Type II lethality	Possible	ND	ND	6
N. ingulba	20	Type II lethality	Possible	Q chromosome	ND	6, 10
N. maritima	16	Type II lethality	Possible	Q chromosome	ND	7
N. megalosiphon	20	Type II lethality	Possible	Q chromosome	ND	7
N. occidentalis	21	Type V lethaliy	Possible	S and T subgenomes	ND	5
N. rosulata	20	Type II lethality	Possible	ND	ND	6
N. rotundifolia	22	Type II lethality	Possible	ND	ND	6
N. simulans	20	Type II lethality	Possible	ND	ND	6
N. suaveolens	16	Type II lethality	Possible	Q chromosome	ND	2
N. umbratica	23	Type II lethality	Possible	ND	ND	6
N. velutina	16	Type II lethality	Possible	Q chromosome	ND	7
N. wuttkei	16	Type II lethality	ND	ND	ND	9

[a] ND, not determined

[b] 1, DeVerna et al. (1987); 2, Tezuka & Marubashi (2006b); 3, Tezuka et al. (2006); 4, Tezuka et al. (2007); 5, Tezuka et al. (2009); 6, Iizuka et al. (2010); 7, Tezuka et al. (2010); 8, Iizuka et al. (2011); 9, Laskowska & Berbeć (2011); 10, Matsuo et al. (2011)

Table 1. Hybrid lethality observed in crosses between *Suaveolentes* species and *N. tabacum*

4.1.1 Causative genes in *N. tabacum*

Nicotiana tabacum (2n = 48, SSTT) is a natural allotetraploid (amphidiploid) that originated by interspecific hybridization of *N. sylvestris* (2n = 24, SS; section *Sylvestres*) with *N. tomentosiformis* (2n = 24, TT; section *Tomentosae*) and subsequent chromosome doubling (Chase et al., 2003; Clarkson et al., 2004, 2010; Gray et al., 1974; Lim et al., 2000; Murad et al., 2002; Sheen, 1972). Therefore, it is possible to determine which subgenome of *N. tabacum* is involved in hybrid lethality using these progenitors.

Nicotiana debneyi (Tezuka et al., 2007) and *N. suaveolens* (Inoue et al., 1996) were crossed with the two progenitors of *N. tabacum*. Both *N. debneyi* and *N. suaveolens* produced inviable hybrids in crosses with *N. sylvestris*, whereas the two species produced viable hybrids in crosses with *N. tomentosiformis*. These results clearly indicated that the S subgenome of *N. tabacum* is involved in hybrid lethality.

Each chromosome of *N. tabacum* is lettered alphabetically (A–Z, excluding X and Y); chromosomes A–L belong to the T subgenome and M–Z to the S subgenome. A complete set of 24 monosomic lines of *N. tabacum* (Haplo-A–Z), which lack a certain chromosome, has been established in the genetic background of 'Red Russian' (Cameron, 1959; Clausen & Cameron, 1944). These monosomic lines are useful to locate genes on specific chromosomes (Clausen & Cameron, 1944; Kubo et al., 1982).

The first application of *N. tabacum* monosomic lines to study hybrid lethality in *Nicotiana* was in a cross between *N. tabacum* and *N. africana* (Gerstel et al., 1979). When all 24 monosomic lines were crossed with *N. africana*, only Haplo-H produced a high number of viable hybrids. Based on these results, the H chromosome, which belongs to the T subgenome, is considered to be related to hybrid lethality. However, it was not clear whether the viable hybrids from the cross Haplo-H × *N. africana* definitely lacked the H chromosome in their study.

Monosomic lines of *N. tabacum* were used next to investigate hybrid lethality in crosses between *N. tabacum* and *N. suaveolens*. Ten monosomic lines of the S subgenome (Haplo-M–Z, excluding Haplo-P and Haplo-V) were crossed with *N. suaveolens* (Marubashi & Onosato, 2002). A small number of viable hybrids were obtained only from the cross using Haplo-Q. These hybrids possessed 38 or 39 chromosomes, which indicated they lacked the Q chromosome. Therefore, it was speculated that the Q chromosome encodes one or more genes causing hybrid lethality. This was conclusively proven with analyses using Q-chromosome-specific DNA markers (Tezuka & Marubashi, 2006b; Tezuka et al., 2004).

In crosses between the other 10 *Suaveolentes* species and *N. tabacum*, causative gene(s) in *N. tabacum* of hybrid lethality were encoded on the Q chromosome (Table 1; Matsuo et al., 2011; Tezuka et al., 2006, 2007, 2010). These results suggested that many species of section *Suaveolentes* share the same gene(s) that triggers hybrid lethality by interaction with the gene(s) on the Q chromosome.

However, it seems one or more species in section *Suaveolentes* are exceptional with regard to hybrid lethality. *Nicotiana occidentalis* yields inviable hybrids showing Type V lethality in crosses with *N. tabacum* as mentioned already. Hybrid seedlings from this cross combination die despite the lack of the Q chromosome. Based on genetic analyses using the two progenitors of *N. tabacum*, both the S and T subgenomes of *N. tabacum* are apparently related to hybrid lethality in the cross with *N. occidentalis* (Table 1; Tezuka et al., 2009).

4.1.2 Causative genes in *Suaveolentes* species

A segregation analysis by classical Mendelian genetics was conducted to identify the causative genes in *Suaveolentes* species. *Nicotiana debneyi* and *N. fragrans* yield inviable and viable hybrids, respectively, in crosses with *N. tabacum* (Table 1; Tezuka et al., 2007, 2010). F_1 hybrids obtained from the cross *N. debneyi* × *N. fragrans* were crossed with *N. tabacum* (Iizuka et al., 2011). Trispecific hybrids from this cross were segregated into inviable and viable hybrids with a ratio of 1:1. Therefore, it was determined that *N. debneyi* carries a single dominant gene causing hybrid lethality in the cross with *N. tabacum*. The gene locus was designated *HYBRID LETHALITY A1* (*HLA1*) and the *N. debneyi* allele, which causes hybrid lethality, was assigned as *Hla1-1* and the non-causative allele of *N. fragrans* and *N. tabacum* as *hla1-2* (Iizuka et al., 2011). In conclusion, the *Hla1-1* allele in *N. debneyi* causes hybrid lethality by interaction with the gene(s) on the Q chromosome of *N. tabacum*.

Similar to hybrid lethality in the cross using *N. debneyi*, gene(s) on the Q chromosome cause Type II lethality in other crosses involving 10 *Suaveolentes* species (Table 1; Matsuo et al., 2011; Tezuka & Marubashi, 2006b; Tezuka et al., 2006, 2010). These results infer that the *Hla1-1* allele is shared by at least 11 *Suaveolentes* species (*N. africana*, *N. amplexicaulis*, *N. debneyi*, *N. excelsior*, *N. goodspeedii*, *N. gossei*, *N. ingulba*, *N. maritima*, *N. megalosiphon*, *N. suaveolens* and *N. velutina*). This finding is consistent with the hypothesis that section *Suaveolentes* originated from a single polyploid event some 10 Mya, followed by speciation to produce the extant species (Leitch et al., 2008). This hypothesis is supported by recent studies that indicate section *Suaveolentes* is a monophyletic group based on sequence data for the internal transcribed spacer region (Chase et al., 2003), plastid genes (Clarkson et al., 2004) and nuclear-encoded chloroplast-expressed glutamine synthetase (ncpGS; Clarkson et al., 2010).

All species in section *Suaveolentes* are allotetraploids. This section contains an almost complete aneuploid series of $n = 16$–24, with only $n = 17$ unknown. According to Clarkson et al. (2004), it seems that the allotetraploid ancestor of the section occurred in South America, where its parental species are found, and subsequently the allotetraploid ancestor dispersed to Africa and Australia separately; only in Australia has an explosive radiation of taxa occurred, largely accompanied by dysploid reductions probably because of chromosomal fusions. Progenitors (parental species of the allotetraploid ancestor) of this section have been proposed by some researchers. Goodspeed (1954) considered that, based on external morphology, ancestral races related to the source of the sections *Alatae*, *Noctiflorae* and *Petunioides* were involved in the formation of section *Suaveolentes*. However, based on analyses of ncpGS sequences, Clarkson et al. (2010) recently suggested the maternal progenitor is *N. sylvestris* and the paternal progenitor is section *Trigonophyllae* ($2n = 24$). Although the evidence for the progenitors of section *Suaveolentes* is inconclusive at present, this information might shed light on the origin of the *Hla1-1* allele. *Nicotiana sylvestris* produces viable hybrids in reciprocal crosses with *N. tabacum* (Christoff, 1928; East, 1935; Kostoff, 1930; Tanaka, 1961). Therefore, the *Hla1-1* allele might be derived from section *Trigonophyllae*. Another possibility is that the allotetraploid ancestor of section *Suaveolentes* or the descendant of the allotetraploid ancestor acquired the *Hla1-1* allele after divergence of *N. benthamiana* and *N. fragrans*.

As already stated, three *Suaveolentes* species (*N. occidentalis*, *N. benthamiana* and *N. fragrans*) yielded results different from those using the above-mentioned 11 species. *Nicotiana benthamiana* and *N. fragrans*, which yield viable hybrids in crosses with *N. tabacum*, possess the *hla1-2* allele. The causative gene(s) in *N. occidentalis* is somewhat complicated, because this species shows hybrid lethality in which the S and T subgenomes of *N. tabacum* are involved (Table 1; Tezuka et al., 2009). These results suggest genetic changes that reinforce reproductive isolation with *N. tabacum* have accumulated in the lineage leading to *N. occidentalis*. Whether *N. occidentalis* possesses the *Hla1-1* allele is an interesting issue requiring further investigation.

5. Hybrid lethality in other crosses in *Nicotiana*

In addition to crosses between *Suaveolentes* species and *N. tabacum*, hybrid lethality is reported in many interspecific crosses in *Nicotiana*. Some of these crosses involve allotetraploid species as parents, i.e., *N. tabacum* (2n = 48, SSTT) in section *Nicotiana*, *N. rustica* (2n = 48, PPUU) in section *Rusticae*, and species in sections *Polydicliae* (2n = 48), *Repandae* (2n = 48) and *Suaveolentes* (2n = 32–48). In addition to *N. tabacum* and section *Suaveolentes*, recent molecular phylogenetic analysis has revealed the progenitors of these species and sections (Clarkson et al., 2010). The maternal and paternal progenitors of *N. rustica* are, respectively, *N. paniculata* (2n = 24, PP; section *Paniculatae*) and *N. undulata* (2n = 24, UU; section *Undulatae*). Section *Polydicliae* is derived from section *Trigonophyllae* (2n = 24; the maternal progenitor) and *N. attenuata* (2n = 24; section *Petunioides*; the paternal progenitor). Section *Repandae* is derived from *N. sylvestris* (2n = 24, SS; the maternal progenitor) and section *Trigonophyllae* (the paternal progenitor). These progenitors of allotetraploid species might provide important information to determine which subgenome is responsible for hybrid lethality, as in crosses between *N. tabacum* and *Suaveolentes* species.

Tables 2–4 list interspecific crosses using *N. tabacum*, those using *N. rustica*, and other crosses, respectively. Tables 2 and 3 also include crosses using progenitors and relatives of *N. tabacum* and *N. rustica*. In addition to crosses that yield inviable hybrids, those that yield viable hybrids are also listed, because these are useful to discuss hybrid lethality. However, attention was also paid to literature that described viable hybrids. This is because it is likely some authors did not report hybrid lethality, even though they observed this phenomenon, because they placed particular emphasis on the acquisition of viable hybrids that can be used for breeding programs and in studies such as observations of chromosome pairing. For example, Christoff (1928) reported that crosses between *N. tabacum* and *N. alata* produced inviable hybrids that died at different stages of their development, and only two hybrids reached maturity. Nevertheless, Christoff determined this cross as one that produces mature hybrids. Therefore, as for crosses that yield viable hybrids, only articles that describe hybrid lethality in the same study are cited.

In several crosses in Tables 2 and 3, one author reported a certain cross to be inviable but other authors described the same cross as viable. These incongruities might be because of the above-mentioned fact that some authors have not described hybrid lethality in detail. Other explanations of the incongruities might be differences in the parental cultivars or lines used for crosses, or the environmental conditions under which hybrid seedlings were cultivated. Although verification of whether certain cross combinations listed in Tables 2–4 produce viable (or inviable) hybrids might be needed, nonetheless, these data are informative to study and discuss hybrid lethality in *Nicotiana*.

5.1 Hybrid lethality in crosses using *N. tabacum*

Nicotiana tabacum and its progenitors, including closely related species, are reported to yield inviable hybrids in crosses with species in seven *Nicotiana* sections (Table 2). In Table 2, *N. tomentosa* (2n = 24, TT) and *N. otophora* (2n = 24, TT) are included; both species belong to section *Tomentosae* and are closely related to *N. tomentosiformis* (Clarkson et al., 2010). The *N. tabacum* subgenome that encodes gene(s) for hybrid lethality can be estimated in crosses using *N. glutinosa*, which belongs to section *Undulatae*. *Nicotiana glutinosa* × *N. sylvestris* hybrids are viable (Christoff, 1928; East, 1935), but *N. glutinosa* × *N. tomentosiformis* hybrids are inviable (East, 1935; McCray, 1932). Thus, the T subgenome must encode the causative gene(s) for hybrid lethality.

Species[a]		F$_1$ phenotype	References[c]
Female	Male		
Section *Alatae*			
N. tabacum (24)	*N. alata* (9)	Lethality	1, 2, 8
		Viable	3, 6
N. alata (9)	*N. tabacum* (24)	Viable	6
N. tabacum (24)	*N. langsdorffii* (9)	Lethality	2, 8
		Viable	6
N. tabacum (24)	*N. longiflora* (10)	Lethality	1, 2, 6, 8
N. longiflora (10)	*N. tabacum* (24)	Lethality	1, 8
N. tabacum (24)	*N. plumbaginifolia* (10)	Lethality	2, 6, 7
N. tabacum (24)	*N. sanderae* (9)[b]	Viable	2, 3, 6
N. sanderae (9)	*N. tabacum* (24)	Viable	6
Setion *Noctiflorae*			
N. tabacum (24)	*N. glauca* (12)	Viable	3, 6, 8
N. glauca (12)	*N. tabacum* (24)	Viable	3, 6
N. glauca (12)	*N. sylvestris* (12)	Lethality	6
N. tomentosa (12)	*N. glauca* (12)	Viable	3, 6
N. glauca (12)	*N. tomentosa* (12)	Viable	3, 6
N. tomentosiformis (12)	*N. glauca* (12)	Viable	3, 6
N. glauca (12)	*N. tomentosiformis* (12)	Viable	3, 6
Section *Trigonophyllae*			
N. tabacum (24)	*N. trigonophylla* (12)	Viable	8
N. trigonophylla (12)	*N. otophora* (12)	Type III lethality	13
N. trigonophylla (12)	*N. tomentosa* (12)	Viable	4
N. palmeri (12)	*N. tomentosa* (12)	Viable	4
N. palmeri (12)	*N. tomentosiformis* (12)	Viable	4
Section *Undulatae*			
N. tabacum (24)	*N. glutinosa* (12)	Lethality	4, 8
		Viable	2, 4, 6
N. glutinosa (12)	*N. tabacum* (24)	Lethality	8
		Viable	6

Species[a]		F₁ phenotype	References[c]
Female	Male		
Section *Undulatae*			
N. glutinosa (12)	*N. sylvestris* (12)	Viable	2, 6
N. glutinosa (12)	*N. tomentosa* (12)	Lethality	6
N. glutinosa (12)	*N. tomentosiformis* (12)	Lethality	4, 6
Section *Polydicliae*			
N. tabacum (24)	*N. bigelovii* (24)	Viable	6
N. bigelovii (24)	*N. tabacum* (24)	Viable	2, 6
N. tabacum (24)	*N. quadrivalvis* (24)	Viable	6
N. quadrivalvis (24)	*N. tabacum* (24)	Viable	6
N. bigelovii (24)	*N. sylvestris* (12)	Lethality	2
N. tomentosa (12)	*N. bigelovii* (24)	Lethality	6
N. bigelovii (24)	*N. tomentosa* (12)	Lethality	4
		Viable	6
N. tomentosiformis (12)	*N. bigelovii* (24)	Lethality	6
N. bigelovii (24)	*N. tomentosiformis* (12)	Viable	6
N. tomentosa (12)	*N. quadrivalvis* (24)	Lethality	6
N. quadrivalvis (24)	*N. tomentosa* (12)	Viable	6
N. tomentosiformis (12)	*N. quadrivalvis* (24)	Lethality	6
N. quadrivalvis (24)	*N. tomentosiformis* (12)	Viable	6
Section *Repandae*			
N. nesophila (24)	*N. tabacum* (24)	Viable	9
N. nudicaulis (24)	*N. tabacum* (24)	Type I lethality	2, 4, 5, 13
N. tabacum (24)	*N. repanda* (24)	Lethality	11
N. repanda (24)	*N. tabacum* (24)	Type III lethality	9, 10, 11, 12, 13
N. stocktonii (24)	*N. tabacum* (24)	Viable	9
N. nudicaulis (24)	*N. sylvestris* (12)	Lethality	4
N. nudicaulis (24)	*N. tomentosiformis* (12)	Viable	4
N. repanda (24)	*N. sylvestris* (12)	Viable	14
N. repanda (24)	*N. tomentosiformis* (12)	Lethality	14
Sections *Rusticae* and *Paniculatae*			
N. tabacum (24)	*N. rustica* (24)	Viable	3, 6
N. rustica (24)	*N. tabacum* (24)	Lethality	3
		Viable	2, 6, 8
N. tabacum (24)	*N. paniculata* (12)	Viable	8
N. paniculata (12)	*N. tabacum* (24)	Viable	3, 6, 8
N. knightiana (12)	*N. tabacum* (24)	Viable	8

[a] Number in parentheses is the haploid chromosome number
[b] Hybrid taxon, a hybrid between *N. forgetiana* (section *Alatae*) and *N. alata* (Goodspeed, 1954)
[c] 1, Malloch & Malloch (1924); 2, Christoff (1928); 3, Kostoff (1930); 4, McCray (1932); 5, McCray (1933); 6, East (1935); 7, Moav & Cameron (1960); 8, Tanaka (1961); 9, Reed & Collins (1978); 10, Iwai et al. (1985); 11, DeVerna et al. (1987); 12, Shintaku et al. (1988); 13, Yamada et al. (1999); 14, Kobori & Marubashi (2004)

Table 2. Interspecific crosses using *N. tabacum* and its progenitors and closely related species

In crosses using two *Repandae* species, sufficient data are available to determine the *N. tabacum* subgenome that encodes the causative gene(s) for hybrid lethality. Hybrid lethality is observed in the crosses *N. nudicaulis* × *N. tabacum* and *N. nudicaulis* × *N. sylvestris*, but not in the cross *N. nudicaulis* × *N. tomentosiformis* (Christoff, 1928; McCray, 1932, 1933; Yamada et al., 1999). These results indicate the S subgenome encodes the causative gene(s) for hybrid lethality in the cross *N. nudicaulis* × *N. tabacum*. On the other hand, hybrid lethality is observed in the crosses *N. repanda* × *N. tabacum* and *N. repanda* × *N. tomentosiformis*, but not in the cross *N. repanda* × *N. sylvestris* (DeVerna et al., 1987; Kobori & Marubashi, 2004; Iwai et al., 1985; Reed & Collins, 1978; Shintaku et al., 1988), which indicates the T subgenome encodes the causative gene(s) for hybrid lethality in the cross *N. repanda* × *N. tabacum*. Additionally, the types of hybrid lethality differ between the crosses *N. nudicaulis* × *N. tabacum* and *N. repanda* × *N. tabacum* (Yamada et al., 1999). It is interesting that *N. nudicaulis* and *N. repanda*, which are closely related (Chase et al., 2003; Clarkson et al., 2004, 2010), yield different outcomes with regard to hybrid lethality.

Some conflicting results among crosses involving *N. tabacum* and its progenitors are apparent. In crosses with *N. glauca*, which belongs to section *Noctiflorae*, *N. sylvestris* produces inviable hybrids whereas *N. tabacum* produces viable hybrids (East, 1935; Kostoff, 1930; Tanaka, 1961). Similarly, in crosses with *N. bigelovii* and *N. quadrivalvis*, which belong to section *Polydicliae*, *N. sylvestris* and/or *N. tomentosiformis* produce inviable hybrids but *N. tabacum* produces viable hybrids (Christoff, 1928; East, 1935).

5.2 Hybrid lethality in crosses using *N. rustica*

As for *N. tabacum*, *N. rustica* has been crossed with wild tobacco species and reported to produce inviable hybrids (Table 3). In several crosses with *Alatae* species, *N. rustica* and its progenitor, *N. paniculata*, produce inviable hybrids (Christoff, 1928; East, 1935; Kostoff, 1930; Malloch & Malloch, 1924; McCray, 1932; Yamada et al., 1999). These results infer that the causative gene(s) for hybrid lethality is encoded in the P subgenome of *N. rustica*. Involvement of the U subgenome is unclear, because crossing results using another progenitor, *N. undulata*, are not available.

The causative gene(s) for hybrid lethality in the cross *N. rustica* × *N. suaveolens* would be also encoded in the P subgenome, because hybrid lethality is observed in the cross *N. paniculata* × *N. suaveolens* (Christoff, 1928; East, 1935; Yamada et al., 1999). *Nicotiana palmeri*, which belongs to section *Trigonophyllae*, produces inviable hybrids in crosses with *N. rustica* (Kostoff, 1930). Therefore, the causative gene(s) for hybrid lethality in *N. nudicaulis*, *N. suaveolens* and *N. gossei* might have been derived from section *Trigonophyllae*, although involvement of *N. sylvestris* cannot be ruled out.

5.3 Hybrid lethality in other crosses

Table 4 shows results of crosses using species other than *N. tabacum*, *N. rustica* and their relatives. Noteworthy is hybrid lethality in crosses between *Alatae* species and *Polydicliae* or *Suaveolentes* species. Sections *Polydicliae* and *Suaveolentes* share the same progenitor, i.e., section *Trigonophyllae* (Clarkson et al., 2010). *Nicotiana trigonophylla* in section *Trigonophyllae* produces inviable hybrids in crosses with *N. langsdorffii* from section *Alatae* (Christoff, 1928). Therefore, at least in the crosses *N. bigelovii* × *N. langsdorffii* and *N. suaveolens* × *N. langsdorffii*, the causative gene(s) for hybrid lethality in *N. bigelovii* and *N. suaveolens* might have been derived from section *Trigonophyllae*. Nonetheless, gene(s) derived from other progenitors (*N. attenuata* for *N. bigelovii* and *N. sylvestris* for *N. suaveolens*) might be involved in hybrid lethality.

Species[a]		F_1 phenotype	References[c]
Female	Male		
Section *Alatae*			
N. rustica (24)	*N. alata* (9)	Lethality	3
		Viable	5
N. rustica (24)	*N. langsdorffii* (9)	Lethality	1, 3, 4
		Viable	2, 5
N. rustica (24)	*N. sanderae* (9)[b]	Lethality	3
		Viable	2, 5
N. paniculata (12)	*N. alata* (9)	Type IV lethality	2, 6
		Viable	5
N. alata (9)	*N. paniculata* (12)	Lethality	5
N. paniculata (12)	*N. langsdorffii* (9)	Lethality	3, 5
		Viable	2
N. paniculata (12)	*N. longiflora* (10)	Lethality	2, 5
N. paniculata (12)	*N. plumbaginifolia* (10)	Lethality	2, 5
N. plumbaginifolia (10)	*N. paniculata* (12)	Lethality	2
N. paniculata (12)	*N. sanderae* (9)	Viable	2, 5
N. sanderae (9)	*N. paniculata* (12)	Lethality	5
Section *Petunioides*			
N. rustica (24)	*N. attenuata* (12)	Lethality	3
Section *Trigonophyllae*			
N. rustica (24)	*N. palmeri* (12)	Lethality	3
Section *Undulatae*			
N. paniculata (12)	*N. glutinosa* (12)	Type IV lethality	6
		Viable	2, 5
N. glutinosa (12)	*N. paniculata* (12)	Lethality	5
Section *Polydicliae*			
N. rustica (24)	*N. bigelovii* (24)	Viable	5
N. rustica (24)	*N. quadrivalvis* (24)	Viable	5
N. bigelovii (24)	*N. paniculata* (12)	Lethality	2
N. paniculata (12)	*N. bigelovii* (24)	Viable	5
N. paniculata (12)	*N. quadrivalvis* (24)	Viable	5
Section *Repandae*			
N. paniculata (12)	*N. nudicaulis* (24)	Type III lethality	6
Section *Suaveolentes*			
N. rustica (24)	*N. suaveolens* (16)	Lethality	2
N. paniculata (12)	*N. gossei* (18)	Type II lethality	6
N. paniculata (12)	*N. suaveolens* (16)	Type II lethality	2, 5, 6

[a] Number in parentheses is the haploid chromosome number
[b] Hybrid taxon, a hybrid between *N. forgetiana* (section *Alatae*) and *N. alata* (Goodspeed, 1954)
[c] 1, Malloch & Malloch (1924); 2, Christoff (1928); 3, Kostoff (1930); 4, McCray (1932); 5, East (1935); 6, Yamada et al. (1999)

Table 3. Interspecific crosses using *N. rustica* and its progenitors

Species[a]		F₁ phenotype	References[c]
Female	Male		
Sections *Alatae* and *Trigonophyllae*			
N. trigonophylla (12)	*N. langsdorffii* (9)	Lethality	2
Sections *Alatae* and *Undulatae*			
N. glutinosa (12)	*N. langsdorffii* (9)	Lethality	2, 4
Sections *Alatae* and *Polydicliae*			
N. bigelovii (24)	*N. langsdorffii* (9)	Lethality	2
N. bigelovii (24)	*N. longiflora* (10)	Lethality	1, 2
N. bigelovii (24)	*N. plumbaginifolia* (10)	Lethality	2
N. bigelovii (24)	*N. sanderae* (9)[b]	Lethality	1
Sections *Alatae* and *Suaveolentes*			
N. alata (9)	*N. suaveolens* (16)	Lethality	4
N. suaveolens (16)	*N. alata* (9)	Lethality	2, 4
N. gossei (18)	*N. alata* (9)	Type IV lethality	5
N. suaveolens (16)	*N. langsdorffii* (9)	Lethality	4
N. megalosiphon (20)	*N. longiflora* (10)	Type III lethality	5
N. suaveolens (16)	*N. longiflora* (10)	Viable	2, 3, 4
N. megalosiphon (20)	*N. plumbaginifolia* (10)	Type III lethality	5
N. suaveolens (16)	*N. plumbaginifolia* (10)	Viable	2, 4
N. sanderae (9)	*N. suaveolens* (16)	Lethality	4
N. suaveolens (16)	*N. sanderae* (9)	Lethality	4
Sections *Repandae* and *Undulatae*			
N. glutinosa (12)	*N. nudicaulis* (24)	Viable	2
N. glutinosa (12)	*N. repanda* (24)	Type III lethality	5
Sections *Repandae* and *Suaveolentes*			
N. debneyi (24)	*N. repanda* (24)	Type II lethality	5

[a] Number in parentheses is the haploid chromosome number
[b] Hybrid taxon, a hybrid between *N. forgetiana* (section *Alatae*) and *N. alata* (Goodspeed, 1954)
[c] 1, Malloch & Malloch (1924); 2, Christoff (1928); 3, Kostoff (1930); 4, East (1935); 5, Yamada et al. (1999)

Table 4. Other interspecific crosses in *Nicotiana*

The cross between *N. debneyi* from section *Suaveolentes* and *N. repanda* from section *Repandae* produces inviable hybrids (Yamada et al., 1999). These sections are derived from shared progenitors, i.e., *N. sylvestris* and section *Trigonophyllae* (Clarkson et al., 2010). When *N. debneyi* and *N. repanda* are crossed with *N. sylvestris*, the former produces inviable hybrids (Tezuka et al., 2007), whereas the latter produces viable hybrids (Kobori & Marubashi, 2004). Therefore, the causative gene in *N. repanda* might have been derived from *N. sylvestris*.

6. Methods to overcome hybrid lethality

For plant breeders, it is important to overcome hybrid lethality, since this mechanism can be an obstacle when seeking to introduce desirable genes into cultivated species. If hybrid lethality can be overcome, a larger germplasm pool will be available in breeding programs. Several methods to overcome or suppress hybrid lethality have been developed through studies in the genus *Nicotiana*.

6.1 Cultivation at elevated temperatures

Hybrid seedlings show hybrid lethality at 28°C, a temperature suitable for the growth of tobacco plants. When hybrid seedlings are cultivated at elevated temperatures generally ranging from 32–38°C, the seedlings may grow normally without exhibiting lethal symptoms. Following the first report of temperature sensitivity in the cross *N. suaveolens* × *N. tabacum* (Manabe et al., 1989), hybrid lethality has been demonstrated to be suppressed at elevated temperatures in many *Nicotiana* crosses (Iizuka et al., 2010; Marubashi & Kobayashi, 2002; Mino et al., 2002; Tezuka et al., 2006, 2010; Watanabe & Marubashi, 2004; Yamada et al., 1999). Obtaining even hybrid seedlings that grow to maturity and flower is possible by suppressing hybrid lethality at elevated temperatures at least in some crosses (Manabe et al., 1989; Tezuka & Marubashi, 2006a; Yamada et al., 1999). This method of cultivation at elevated temperatures is very simple and convenient. However, hybrid seedlings must be cultivated continuously at elevated temperatures from germination to maturity. When hybrid seedlings are transferred from an elevated temperature to one below 28°C, they will die. In *Nicotiana*, this method is useful for Types I–III and V lethality, but might not be effective against Type IV lethality (Tezuka et al., 2009; Yamada et al., 1999). Temperature sensitivity is also observed in hybrid lethality in other plant species, including intraspecific and interspecific crosses in *Gossypium* (Phillips, 1977), *Oryza* (Saito et al., 2007), *Lactuca* (Jeuken et al., 2009) and *Arabidopsis* (Bomblies et al., 2007), and intergeneric hybrids between *Pyrus pyrifolia* and *Malus* × *domestica* (Inoue et al., 2003). However, why hybrid lethality is suppressed at elevated temperatures is still unclear.

6.2 Tissue culture

Cotyledon culture has been used to overcome hybrid lethality in *Nicotiana* (DeVerna et al., 1987; Lloyd, 1975; Ternovskii et al., 1976; Yamada et al., 1999). To raise viable hybrids with this procedure, cotyledons are excised from hybrid seedlings before showing lethal phenotypes. The cotyledons are sectioned and cultured on an appropriate medium containing auxin and cytokinin to induce callus. Adventitious shoots regenerated from the callus are normal and can be rooted. These plants grow to maturity after acclimatization. Although cotyledons are often used as explant sources, other tissues can be suitable. In the cross *N. occidentalis* × *N. tabacum*, germinated seeds were used for callus and shoot production (Ternovskii et al., 1972). Small leaves were used in the cross *N. repanda* × *N. tabacum* (Iwai et al., 1985). In *Triticum*, viable hybrids were obtained by regeneration from callus induced by culture of immature embryos (Chen et al., 1989).

In *Nicotiana*, cotyledon culture is reported to be useful to overcome Type II lethality, but not Types I, III and IV lethality (Yamada et al., 1999). Nevertheless, Type III lethality in the cross *N. repanda* × *N. tabacum* was overcome by tissue culture (Iwai et al., 1985). Tissue culture is also useful for Type V lethality in the cross *N. occidentalis* × *N. tabacum* (Ternovskii et al., 1972).

The mechanism of overcoming hybrid lethality by tissue culture may be explained by somaclonal variation that is often observed in regenerated plants (Larkin & Scowcroft, 1981). For example, the phenomenon might involve (1) deletion of a chromosome or chromosome segment, leading to deletion of the causative gene for hybrid lethality, (2) mutation in the causative gene(s), leading to loss of function in the induction of hybrid lethality, and (3) mutation in other key genes required to induce hybrid lethality.

6.3 Application of cytokinin or auxin

Application of exogenous plant growth regulators, such as auxin and cytokinin, is reported to be effective to address hybrid lethality. Hybrid seedlings from reciprocal crosses between *N. repanda* and *N. tabacum* cultivated in vermiculite supplemented with 1/4 Murashige and Skoog (MS) solution (Murashige & Skoog, 1962) and the auxin indole-3-acetic acid (IAA) grew to maturity, whereas hybrid seedlings supplied with only 1/4 MS solution died (Zhou et al., 1991). This IAA treatment needed to be continued, otherwise the hybrid seedlings died. The authors suggested the hybrid seedlings contained insufficient endogenous auxin for normal growth and that exogenous IAA compensated for this deficiency. Effectiveness of the IAA treatment in other lethal crosses is unclear.

Inoue et al. (1994) developed a method to overcome hybrid lethality in the cross *N. suaveolens* × *N. tabacum* by cytokinin treatment. This method was carried out in two steps. First, the seeds were germinated on 1/2 MS solid medium. Next, the hybrid seedlings were transplanted to 1/2 MS solid medium containing cytokinin. After this treatment, the hybrid seedlings began to die. However, adventitious buds developed from the base of the hypocotyl and the primary root of the dying seedlings, and grew into normal green shoots that were viable without exogenous cytokinin. After rooting, the plantlets grew to maturity. This two-step procedure was further improved by the same authors; hybrid seeds were cultured in 1/2 MS liquid medium containing cytokinin (Inoue et al., 1997). This simple one-step method allowed the whole surface of hybrid seedlings to absorb cytokinin. Adventitious shoots were regenerated from the root and the whole hypocotyl, and at a greater frequency compared with the two-step method. In both methods, several types of cytokinin, including kinetin, 6-benzylaminopurine, *t*-zeatin, thidiazuron and *N*-(2-chloro-4-pyridyl)-*N'*-phenylurea, were effective to overcome hybrid lethality. However, the efficiency of shoot regeneration differed with the type and concentration of cytokinin (Inoue et al., 1994, 1997).

Although the mechanism of overcoming hybrid lethality by exogenous cytokinin application is still unclear, three hypotheses were proposed by Inoue et al. (1997): (1) cytokinin directly suppresses hybrid lethality, (2) cytokinin induces a mutation in the causative gene(s) for hybrid lethality, and (3) cytokinin enables the screening of variant cells that carry a spontaneous mutation in the causative gene(s).

In *Nicotiana*, cytokinin culture methods are effective for Types I and II lethality, but might not overcome Types III and IV lethality (Yamada et al., 1999). Effectiveness in other plant species is unclear.

6.4 Irradiation

Irradiation with γ-rays and ion beams has been used to overcome hybrid lethality. Irradiation with γ-rays was used to overcome Type III lethality in a cross between *N. repanda* and *N. tabacum*. Viable hybrids were obtained when *N. repanda* was pollinated with γ-ray-irradiated pollen of *N. tabacum*, and when γ-ray-irradiated egg cells (ovules) of *N. repanda* were fertilized by *N. tabacum* (Shintaku et al., 1988, 1989). To overcome Type II lethality in the cross *N. gossei* × *N. tabacum*, *N. tabacum* pollen irradiated with γ-rays or ion beams was used successfully (Kitamura et al., 2003).

Irradiation techniques are also effective in other plant species. In *Triticum*, chimera hybrids, which produced necrotic and normal tillers, and necrotic and normal leaves, were obtained by irradiation of hybrid seeds with γ-rays. These chimeras set seeds (Sharma, 1969). In the intergeneric cross between *Pyrus pyrifolia* and *Malus* × *domestica*, hybrids without lethal phenotypes were obtained by γ-ray irradiation of shoots from immature hybrid embryos (Gonai et al., 2006). Although direct evidence is not available, irradiation might allow elimination or mutation of the causative gene(s) or other key genes that contribute to hybrid lethality.

7. Conclusion

The genus *Nicotiana* is advantageous for investigation of hybrid lethality because many cross combinations show hybrid lethality. Several gene combinations probably cause hybrid lethality in *Nicotiana*. However, as yet none of the genes have been cloned and characterized. Recently, the causative gene for hybrid lethality in *Arabidopsis thaliana* was identified (Bomblies et al., 2007). In intraspecific crosses in *A. thaliana*, hybrid lethality is induced by the interaction between an allele of the *DANGEROUS MIX 1* (*DM1*) locus and an allele of the *DM2* locus. *DM1* encodes a *Toll Interleukin Receptor-Nucleotide Binding-Leucine Rich Repeat* (*TIR-NB-LRR*) disease resistance (*R*) gene homolog. This finding strongly indicates an autoimmune-like response is the mechanism for hybrid lethality in *A. thaliana* (Bomblies et al., 2007).

In interspecific crosses of lettuce (*Lactuca saligna* × *L. sativa*), a specific allelic combination at two loci triggers hybrid lethality. One of the two interacting loci was *Rin4*, a homolog of *RPM1 INTERACTING PROTEIN 4* (*RIN4*) of *A. thaliana* (Jeuken et al., 2009). In *A. thaliana*, RIN4 is a target of at least three effectors (AvrB, AvrRpm1 and AvrRpt2) from *Pseudomonas syringae*, and is guarded by two R proteins, RPM1 and RPS2 (Axtell & Staskawicz, 2003; Kim et al., 2005; Mackey et al., 2002, 2003). Interaction of RIN4 and these R proteins results in the hypersensitive response to *P. syringae*. Similarly, association between hybrid lethality and disease resistance has been shown in interspecific crosses involving tomato. The *Cf-2* gene confers resistance to *Cladosporium fulvum*. When a dominant allele of *Cf-2* derived from *Solanum pimpinellifolium* is combined with a recessive allele of *RCR3* from *S. lycopersicum*, autonecrosis is observed (Day, 1958; Krüger et al., 2002; Langford, 1948; Santangelo et al., 2003). This phenomenon infers hybrid lethality, or more strictly hybrid breakdown. *Cf-2* encodes a transmembrane protein with an extracellular LRR and *RCR3* encodes a secreted papain-like cysteine endoprotease (Dixon et al., 1996; Krüger et al., 2002).

Thus, recent findings have revealed that the mechanism of hybrid lethality is related to disease resistance. However, whether all types of hybrid lethality are explained by disease resistance is uncertain. To address this question, myself and co-workers plan to elucidate the distribution of the causative genes, conduct genetic analyses, and identify and clone the causative genes in *Nicotiana*. These studies will reveal the diverse mechanisms of hybrid lethality, contribute to the development of new cultivars and also help to understand speciation mechanisms.

8. Acknowledgment

Preparation of this chapter was partly supported by a Grant-in-Aid for Young Scientists (Start-up) No. 20880024 from the Japan Society for the Promotion of Science.

9. References

Axtell, M.J. & Staskawicz, B.J. (2003). Initiation of *RPS2*-specified disease resistance in *Arabidopsis* is coupled to the AvrRpt2-directed elimination of RIN4. *Cell*, Vol.112, pp. 369-377, ISSN 0092-8674

Bai, D.; Reeleder, R. & Brandle, J.E. (1995). Identification of two RAPD markers tightly linked with the *Nicotiana debneyi* gene for resistance to black root rot of tobacco. *Theoretical and Applied Genetics*, Vol.91, pp. 1184-1189, ISSN 0040-5752

Bomblies, K.; Lempe, J.; Epple, P.; Warthmann, N.; Lanz, C.; Dangl, J.L. & Weigel, D. (2007). Autoimmune response as a mechanism for a Dobzhansky-Muller-type incompatibility syndrome in plants. *PLoS Biology*, Vol.5, e236, ISSN 1544-9173

Bomblies, K. & Weigel, D. (2007). Hybrid necrosis: autoimmunity as a potential gene-flow barrier in plant species. *Nature Reviews Genetics*, Vol.8, pp. 382-393, ISSN 1471-0056

Burk, L.G. & Heggestad, H.E. (1966). The genus *Nicotiana*: a source of resistance to diseases of cultivated tobacco. *Economic Botany*, Vol.20, pp. 76-88, ISSN 0013-0001

Cameron, D.R. (1959). The monosomics of *Nicotiana tabacum*. *Tobacco Science*, Vol.3, pp. 164-166

Chase, M.W.; Knapp, S.; Cox, A.V.; Clarkson, J.J.; Butsko, Y.; Joseph, J.; Savolainen, V. & Parokonny, A.S. (2003). Molecular systematics, GISH and the origin of hybrid taxa in *Nicotiana* (Solanaceae). *Annals of Botany*, Vol.92, pp. 107-127, ISSN 0305-7364

Chen, Z.; Evans, D.A. & Vasconcelos, A. (1989). Use of tissue culture to bypass wheat hybrid necrosis. *Theoretical and Applied Genetics*, Vol.78, pp. 57-60, ISSN 0040-5752

Christoff, M. (1928). Cytological studies in the genus *Nicotiana*. *Genetics*, Vol.13, pp. 233-277, ISSN 0016-6731

Clarkson, J.J.; Knapp, S.; Garcia, V.F.; Olmstead, R.G.; Leitch, A.R. & Chase, M.W. (2004). Phylogenetic relationships in *Nicotiana* (Solanaceae) inferred from multiple plastid DNA regions. *Molecular Phylogenetics and Evolution*, Vol.33, pp. 75-90, ISSN 1055-7903

Clarkson, J.J.; Kelly, L.J.; Leitch, A.R.; Knapp, S. & Chase, M.W. (2010). Nuclear glutamine synthetase evolution in *Nicotiana*: phylogenetics and the origins of allotetraploid and homoploid (diploid) hybrids. *Molecular Phylogenetics and Evolution*, Vol.55, pp. 99-112, ISSN 1055-7903

Clausen, R.E. & Cameron, D.R. (1944). Inheritance in *Nicotiana tabacum*. XVIII. Monosomic analysis. *Genetics*, Vol.29, pp. 447-477, ISSN 0016-6731

Day, P.R. (1958). Autogenous necrosis in the tomato. *Plant Pathology*, Vol.7, pp. 57-58, ISSN 0032-0862

DeVerna, J.W.; Myers, J.R. & Collins, G.B. (1987). Bypassing prefertilization barriers to hybridization in *Nicotiana* using in vitro pollination and fertilization. *Theoretical and Applied Genetics*, Vol.73, pp. 665-671, ISSN 0040-5752

Dixon, M.S.; Jones, D.A.; Keddie, J.S.; Thomas, C.M.; Harrison, K. & Jones, J.D.G. (1996). The tomato *Cf-2* disease resistance locus comprises two functional genes encoding leucine-rich repeat proteins. *Cell*, Vol.84, pp. 451-459, ISSN 0092-8674

East, E.M. (1935). Genetic reactions in *Nicotiana*. I. Compatibility. *Genetics*, Vol.20, pp. 403-413, ISSN 0016-6731

Gerstel, D.U.; Burns, J.A. & Burk, L.G. (1979). Interspecific hybridizations with an African tobacco, *Nicotiana africana* Merxm. *The Journal of Heredity*, Vol.70, pp. 342-344, ISSN 0022-1503

Gonai, T.; Manabe, T.; Inoue, E.; Hayashi, M.; Yamamoto, T.; Hayashi, T.; Sakuma, F. & Kasumi, M. (2006). Overcoming hybrid lethality in a cross between Japanese pear and apple using gamma irradiation and confirmation of hybrid status using flow cytometry and SSR markers. *Scientia Horticulturae*, Vol.109, pp. 43-47, ISSN 0304-4238

Goodspeed, T.H. (1954). The genus *Nicotiana*. Chronica Botanica Company, Waltham, Massachusetts

Gray, J.C.; Kung, S.D.; Wildman, S.G. & Sheen, S.J. (1974). Origin of *Nicotiana tabacum* L. detected by polypeptide composition of Fraction I protein. *Nature*, Vol.252, pp. 226-227, ISSN 0028-0836

Holmes, F.O. (1938). Inheritance of resistance to tobacco-mosaic disease in tobacco. *Phytopathology*, Vol.28, pp. 553-561, ISSN 0031-949X

Iizuka, T.; Oda, M. & Tezuka, T. (2010). Hybrid lethality expressed in hybrids between *Nicotiana tabacum* and 50 lines of 21 wild species in section *Suaveolentes*. *Breeding Research*, Vol.12 (Suppl. 1), pp. 244, ISSN 1344-7629 (in Japanese)

Iizuka, T.; Kuboyama, T.; Marubashi, W.; Oda, M. & Tezuka T. (2011). *Nicotiana debneyi* has a single dominant gene causing hybrid lethality in the cross with *N. tabacum*. *Breeding Research*, Vol.13 (Suppl. 1), pp. 91, ISSN 1344-7629 (in Japanese)

Inoue, E.; Marubashi, W. & Niwa, M. (1994). Simple method for overcoming the lethality observed in the hybrid between *Nicotiana suaveolens* and *N. tabacum*. *Breeding Science*, Vol.44, pp. 333-336, ISSN 1344-7610

Inoue, E.; Marubashi, W. & Niwa, M. (1996). Genomic factors controlling the lethality exhibited in the hybrid between *Nicotiana suaveolens* Lehm. and *N. tabacum* L. *Theoretical and Applied Genetics*, Vol.93, pp. 341-347, ISSN 0040-5752

Inoue, E.; Marubashi, W. & Niwa, M. (1997). Improvement of the method for overcoming the hybrid lethality between *Nicotiana suaveolens* and *N. tabacum* by culture of F_1 seeds in liquid media containing cytokinins. *Breeding Science*, Vol.47, pp. 211-216, ISSN 1344-7610

Inoue, E.; Sakuma, F.; Kasumi, M.; Hara, H. & Tsukihashi, T. (2003). Effect of high-temperature on suppression of the lethality exhibited in the intergeneric hybrid between Japanese pear (*Pyrus pyrifolia* Nakai) and apple (*Malus × domestica* Borkh.). *Scientia Horticulturae*, Vol.98, pp. 385-396, ISSN 0304-4238

Iwai, S.; Kishi, C.; Nakata, K. & Kubo, S. (1985). Production of a hybrid of *Nicotiana repanda* Willd. × *N. tabacum* L. by ovule culture. *Plant Science*, Vol.41, pp. 175-178, ISSN 0168-9452

Jeuken, M.J.W.; Zhang, N.W.; McHale, L.K.; Pelgrom, K.; den Boer, E.; Lindhout, P.; Michelmore, R.W.; Visser, R.G.F. & Niks, R.E. (2009). *Rin4* causes hybrid necrosis and race-specific resistance in an interspecific lettuce hybrid. *The Plant Cell*, Vol.21, pp. 3368-3378, ISSN 1040-4651

Kim, H.S.; Desveaux, D.; Singer, A.U.; Patel, P.; Sondek, J. & Dangl, J.L. (2005). The *Pseudomonas syringae* effector AvrRpt2 cleaves its C-terminally acylated target, RIN4, from *Arabidopsis* membranes to block RPM1 activation. *Proceedings of the National Academy of Sciences of the United States of America*, Vol.102, pp. 6496-6501, ISSN 0027-8424

Kitamura, S.; Inoue, M.; Ohmido, N.; Fukui, K. & Tanaka, A. (2003). Chromosomal rearrangements in interspecific hybrids between *Nicotiana gossei* Domin and *N.*

tabacum L., obtained by crossing with pollen exposed to helium ion beams or gamma-rays. *Nuclear Instruments and Methods in Physics Research Section B-Beam Interactions with Materials and Atoms*, Vol.206, pp. 548-552, ISSN 0168-583X

Knapp, S.; Chase, M.W. & Clarkson, J.J. (2004). Nomenclatural changes and a new sectional classification in *Nicotiana* (Solanaceae). *Taxon*, Vol.53, pp. 73-82, ISSN 0040-0262

Kobori, S. & Marubashi, W. (2004). Programmed cell death detected in interspecific hybrids of *Nicotiana repanda* × *N. tomentosiformis* expressing hybrid lethality. *Breeding Science*, Vol.54, pp. 347-350, ISSN 1344-7610

Kostoff, D. (1930). Ontogeny, genetics, and cytology of *Nicotiana* hybrids. *Genetica*, Vol.12, pp. 33-139, ISSN 0016-6707

Krüger, J.; Thomas, C.M.; Golstein, C.; Dixon, M.S.; Smoker, M.; Tang, S.; Mulder, L. & Jones, J.D.G. (2002). A tomato cysteine protease required for *Cf*-2-dependent disease resistance and suppression of autonecrosis. *Science*, Vol.296, pp. 744-747, ISSN 0036-8075

Kubo, T.; Sato, M.; Tomita, H. & Kawashima, N. (1982). Identification of the chromosome carrying the gene for *cis*-abienol production by the use of monosomics in *Nicotiana tabacum* L. *Tobacco Science*, Vol.26, pp. 126-128

Langford, A.N. (1948). Autogenous necrosis in tomatoes immune from *Cladosporium fulvum* Cooke. *Canadian Journal of Research Section C*, Vol.26, pp. 35-64, ISSN 0366-7405

Larkin, P.J. & Scowcroft, W.R. (1981). Somaclonal variation – a novel source of variability from cell cultures for plant improvement. *Theoretical and Applied Genetics*, Vol.60, pp. 197-214, ISSN 0040-5752

Laskowska, D. & Berbeć, A. (2011). Production and characterization of amphihaploid hybrids between *Nicotiana wuttkei* Clarkson et Symon and *N. tabacum* L. *Euphytica*, doi: 10.1007/s10681-011-0500-4, ISSN 0014-2336

Leitch, I.J.; Hanson, L.; Lim, K.Y.; Kovarik, A.; Chase, M.W.; Clarkson, J.J. & Leitch, A.R. (2008). The ups and downs of genome size evolution in polyploid species of *Nicotiana* (Solanaceae). *Annals of Botany*, Vol.101, pp. 805-814, ISSN 0305-7364

Li, B.-C.; Bass, W.T. & Cornelius, P.L. (2006). Resistance to tobacco black shank in *Nicotiana* species. *Crop Science*, Vol.46, pp. 554-560, ISSN 0011-183X

Lim, K.Y.; Matyášek, R.; Lichtenstein, C.P. & Leitch, A.R. (2000). Molecular cytogenetic analyses and phylogenetic studies in the *Nicotiana* section Tomentosae. *Chromosoma*, Vol.109, pp. 245-258, ISSN 0009-5915

Lloyd, R. (1975). Tissue culture as a means of circumventing lethality in an interspecific *Nicotiana* hybrid. *Tobacco Science*, Vol.19, pp. 4-6

Mackey, D.; Holt, B.F. 3rd; Wiig, A. & Dangl, J.L. (2002). RIN4 interacts with *Pseudomonas syringae* type III effector molecules and is required for RPM1-mediated resistance in *Arabidopsis*. *Cell*, Vol.108, pp. 743-754, ISSN 0092-8674

Mackey, D.; Belkhadir, Y.; Alonso, J.M.; Ecker, J.R. & Dangl, J.L. (2003). *Arabidopsis* RIN4 is a target of the type III virulence effector AvrRpt2 and modulates RPS2-mediated resistance. *Cell*, Vol.112, pp. 379-389, ISSN 0092-8674

Malloch, W.S. & Malloch, F.W. (1924). Species crosses in *Nicotiana*, with particular reference to *N. longiflora* × *N. Tabacum*, *N. longiflora* × *N. Sanderae*, *N. Tabacum* × *N. glauca*. *Genetics*, Vol.9, pp. 261-291, ISSN 0016-6731

Manabe, T.; Marubashi, W. & Onozawa, Y. (1989). Temperature-dependent conditional lethality in interspecific hybrids between *Nicotiana suaveolens* Lehm. and *N. tabacum*

L. *Proceedings of the 6th International Congress of SABRAO*, pp. 459-462, Tsukuba, Japan, August 1989

Marubashi, W. & Kobayashi, M. (2002). Apoptosis detected in hybrids between *Nicotiana debneyi* and *N. tabacum*. *Breeding Research*, Vol.4, pp. 209-214, ISSN 1344-7629 (in Japanese with English summary)

Marubashi, W. & Onosato, K. (2002). Q chromosome controls the lethality of interspecific hybrids between *Nicotiana tabacum* and *N. suaveolens*. *Breeding Science*, Vol.52, pp. 137-142, ISSN 1344-7610

Matsuo, C.; Iizuka, T.; Marubashi, W.; Oda, M. & Tezuka, T. (2011). Identification of the chromosome responsible for hybrid lethality in interspecific hybrids (*Nicotiana tabacum* × *N. ingulba*) by Q-chromosome-specific SSR markers. *Breeding Research*, Vol.13 (Suppl. 1), pp. 90, ISSN 1344-7629 (in Japanese)

McCray, F.A. (1932). Compatibility of certain *Nicotiana* species. *Genetics*, Vol.17, pp. 621-636, ISSN 0016-6731

McCray, F.A. (1933). Embryo development in *Nicotiana* species hybrids. *Genetics*, Vol.18, pp. 95-110, ISSN 0016-6731

Mino, M.; Maekawa, K.; Ogawa, K.; Yamagishi, H. & Inoue, M. (2002). Cell death process during expression of hybrid lethality in interspecific F_1 hybrid between *Nicotiana gossei* Domin and *Nicotiana tabacum*. *Plant Physiology*, Vol.130, pp. 1776-1787, ISSN 0032-0889

Moav, R. & Cameron, D.R. (1960). Genetic instability in *Nicotiana* hybrids. I. The expression of instability in *N. tabacum* × *N. plumbaginifolia*. *American Journal of Botany*, Vol.47, pp. 87-93, ISSN 0002-9122

Murad, L.; Lim, K.Y.; Christopodulou, V.; Matyasek, R.; Lichtenstein, C.P.; Kovarik, A. & Leitch, A.R. (2002). The origin of tobacco's T genome is traced to a particular lineage within *Nicotiana tomentosiformis* (Solanaceae). *American Journal of Botany*, Vol.89, pp. 921-928, ISSN 0002-9122

Murashige, T. & Skoog, F. (1962). A revised medium for rapid growth and bio assays with tobacco tissue cultures. *Physiologia Plantarum*, Vol.15, pp. 473-497, ISSN 0031-9317

Nikova, V.; Pundeva, R. & Petkova, A. (1999). *Nicotiana tabacum* L. as a source of cytoplasmic male sterility in interspecific cross with *N. alata* Link & Otto. *Euphytica*, Vol.107, pp. 9-12, ISSN 0014-2336

Nikova, V.M. & Zagorska, N.A. (1990). Overcoming hybrid incompatibility between *Nicotiana africana* Merxm. and *N. tabacum* and development of cytoplasmically male sterile tobacco forms. *Plant Cell, Tissue and Organ Culture*, Vol.23, pp. 71-75, ISSN 0167-6857

Nikova, V.M.; Zagorska, N.A. & Pundeva, R.S. (1991). Development of four tobacco cytoplasmic male sterile sources using *in vitro* techniques. *Plant Cell, Tissue and Organ Culture*, Vol.27, pp. 289-295, ISSN 0167-6857

Phillips, L.L. (1977). Interspecific incompatibility in *Gossypium*. IV. Temperature-conditional lethality in hybrids of *G. klotzschianum*. *American Journal of Botany*, Vol.64, pp. 914-915, ISSN 0002-9122

Reed, S.M. & Collins, G.B. (1978). Interspecific hybrids in *Nicotiana* through in vitro culture of fertilized ovules. *The Journal of Heredity*, Vol.69, pp. 311-315, ISSN 0022-1503

Saito, T.; Ichitani, K.; Suzuki, T.; Marubashi, W. & Kuboyama, T. (2007). Developmental observation and high temperature rescue from hybrid weakness in a cross between

Japanese rice cultivars and Peruvian rice cultivar 'Jamaica'. *Breeding Science*, Vol.57, pp. 281-288, ISSN 1344-7610

Santangelo, E.; Fonzo, V.; Astolfi, S.; Zuchi, S.; Caccia, R.; Mosconi, P.; Mazzucato, A. & Soressi, G.P. (2003). The *Cf-2/Rcr3esc* gene interaction in tomato (*Lycopersicon esculentum*) induces autonecrosis and triggers biochemical markers of oxidative burst at cellular level. *Functional Plant Biology*, Vol.30, pp. 1117-1125, ISSN 1445-4408

Sharma, D. (1969). Use of radiations for breaking hybrid necrosis in wheat. *Euphytica*, Vol.18, pp. 66-70, ISSN 0014-2336

Sheen, S.J. (1972). Isozymic evidence bearing on the origin of *Nicotiana tabacum* L. *Evolution*, Vol.26, pp. 143-154, ISSN 0014-3820

Shintaku, Y.; Yamamoto, K. & Nakajima, T. (1988). Interspecific hybridization between *Nicotiana repanda* Willd. and *N. tabacum* L. through the pollen irradiation technique and the egg cell irradiation technique. *Theoretical and Applied Genetics*, Vol.76, pp. 293-298, ISSN 0040-5752

Shintaku, Y.; Yamamoto, K. & Takeda, G. (1989). Chromosomal variation in hybrids between *Nicotiana repanda* Willd. and *N. tabacum* L. through pollen and egg-cell irradiation techniques. *Genome*, Vol.32, pp. 251-256, ISSN 0831-2796

Stavely, J.R.; Pittarelli, G.W. & Burk, L.G. (1973). *Nicotiana repanda* as a potential source for disease resistance in *N. tabacum*. *The Journal of Heredity*, Vol.64, pp. 265-271, ISSN 0022-1503

Tanaka, M. (1961). The effect of irradiated pollen grains on species crosses of *Nicotiana*. *Bulletin of the Hatano Tobacco Experiment Station*, Vol.51, pp. 1-38, ISSN 0367-6323 (in Japanese with English summary)

Ternovskii, M.F.; Butenko, R.G. & Moiseeva, M.E. (1972). The use of tissue culture to overcome the barrier of incompatibility between species and sterility of interspecies hybrids. *Soviet Genetics*, Vol.8, pp. 27-33, ISSN 0038-5409, Translated from *Genetika* Vol.8, pp. 38-45, ISSN 0534-0012

Ternovskii, M.F.; Shinkareva, I.K. & Lar'kina, N.I. (1976). Production of interspecific tobacco hybrids by the pollination of ovules in vitro. *Soviet Genetics*, Vol.12, pp. 1209-1213, ISSN 0038-5409, Translated from *Genetika* Vol.12, pp. 40-45, ISSN 0534-0012

Tezuka, T. & Marubashi, W. (2004). Apoptotic cell death observed during the expression of hybrid lethality in interspecific hybrids between *Nicotiana tabacum* and *N. suaveolens*. *Breeding Science*, Vol.54, pp. 59-66, ISSN 1344-7610

Tezuka, T.; Onosato, K.; Hijishita, S. & Marubashi, W. (2004). Development of Q-chromosome-specific DNA markers in tobacco and their use for identification of a tobacco monosomic line. *Plant and Cell Physiology*, Vol.45, pp. 1863-1869, ISSN 0032-0781

Tezuka, T. & Marubashi, W. (2006a). Genomic factors lead to programmed cell death during hybrid lethality in interspecific hybrids between *Nicotiana tabacum* and *N. debneyi*. *SABRAO Journal of Breeding and Genetics*, Vol.38, pp. 69-81, ISSN 1029-7073

Tezuka, T. & Marubashi, W. (2006b). Hybrid lethality in interspecific hybrids between *Nicotiana tabacum* and *N. suaveolens*: evidence that the Q chromosome causes hybrid lethality based on Q-chromosome-specific DNA markers. *Theoretical and Applied Genetics*, Vol.112, pp. 1172-1178, ISSN 0040-5752

Tezuka, T.; Kuboyama, T.; Matsuda, T. & Marubashi, W. (2006). Expression of hybrid lethality in interspecific crosses between *Nicotiana tabacum* and nine wild species of section *Suaveolentes*, and the chromosome responsible for hybrid lethality. *Breeding Research*, Vol.8 (Suppl. 2), pp. 139, ISSN 1344-7629 (in Japanese)

Tezuka, T.; Kuboyama, T.; Matsuda, T. & Marubashi, W. (2007). Possible involvement of genes on the Q chromosome of *Nicotiana tabacum* in expression of hybrid lethality and programmed cell death during interspecific hybridization to *Nicotiana debneyi*. *Planta*, Vol.226, pp. 753-764, ISSN 0032-0935

Tezuka, T.; Oda, M. & Marubashi, W. (2009). Identification of genomes involved in hybrid lethality in interspecific hybrids between *Nicotiana tabacum* and *N. occidentalis*. *Breeding Research*, Vol.11 (Suppl. 1), pp. 314, ISSN 1344-7629 (in Japanese)

Tezuka, T.; Kuboyama, T.; Matsuda, T. & Marubashi, W. (2010). Seven of eight species in *Nicotiana* section *Suaveolentes* have common factors leading to hybrid lethality in crosses with *Nicotiana tabacum*. *Annals of Botany*, Vol.106, pp. 267-276, ISSN 0305-7364

Watanabe, H. & Marubashi, W. (2004). Temperature-dependent programmed cell death detected in hybrids between *Nicotiana langsdorffii* and *N. tabacum* expressing lethality. *Plant Biotechnology*, Vol.21, pp. 151-154, ISSN 1342-4580

Yamada, T.; Marubashi, W. & Niwa, M. (1999). Detection of four lethality types in interspecific crosses among *Nicotiana* species through the use of three rescue methods for lethality. *Breeding Science*, Vol.49, pp. 203-210, ISSN 1344-7610

Zhou, W.M.; Yoshida, K.; Shintaku, Y. & Takeda, G. (1991). The use of IAA to overcome interspecific hybrid inviability in reciprocal crosses between *Nicotiana tabacum* L. and *N. repanda* Willd. *Theoretical and Applied Genetics*, Vol.82, pp. 657-661, ISSN 0040-5752

Permissions

The contributors of this book come from diverse backgrounds, making this book a truly international effort. This book will bring forth new frontiers with its revolutionizing research information and detailed analysis of the nascent developments around the world.

We would like to thank Dr. John Kiogora Mworia, for lending his expertise to make the book truly unique. He has played a crucial role in the development of this book. Without his invaluable contribution this book wouldn't have been possible. He has made vital efforts to compile up to date information on the varied aspects of this subject to make this book a valuable addition to the collection of many professionals and students.

This book was conceptualized with the vision of imparting up-to-date information and advanced data in this field. To ensure the same, a matchless editorial board was set up. Every individual on the board went through rigorous rounds of assessment to prove their worth. After which they invested a large part of their time researching and compiling the most relevant data for our readers. Conferences and sessions were held from time to time between the editorial board and the contributing authors to present the data in the most comprehensible form. The editorial team has worked tirelessly to provide valuable and valid information to help people across the globe.

Every chapter published in this book has been scrutinized by our experts. Their significance has been extensively debated. The topics covered herein carry significant findings which will fuel the growth of the discipline. They may even be implemented as practical applications or may be referred to as a beginning point for another development. Chapters in this book were first published by InTech; hereby published with permission under the Creative Commons Attribution License or equivalent.

The editorial board has been involved in producing this book since its inception. They have spent rigorous hours researching and exploring the diverse topics which have resulted in the successful publishing of this book. They have passed on their knowledge of decades through this book. To expedite this challenging task, the publisher supported the team at every step. A small team of assistant editors was also appointed to further simplify the editing procedure and attain best results for the readers.

Our editorial team has been hand-picked from every corner of the world. Their multi-ethnicity adds dynamic inputs to the discussions which result in innovative outcomes. These outcomes are then further discussed with the researchers and contributors who give their valuable feedback and opinion regarding the same. The feedback is then collaborated with the researches and they are edited in a comprehensive manner to aid the understanding of the subject.

Apart from the editorial board, the designing team has also invested a significant amount of their time in understanding the subject and creating the most relevant covers. They scrutinized every image to scout for the most suitable representation of the subject and create an appropriate cover for the book.

The publishing team has been involved in this book since its early stages. They were actively engaged in every process, be it collecting the data, connecting with the contributors or procuring relevant information. The team has been an ardent support to the editorial, designing and production team. Their endless efforts to recruit the best for this project, has resulted in the accomplishment of this book. They are a veteran in the field of academics and their pool of knowledge is as vast as their experience in printing. Their expertise and guidance has proved useful at every step. Their uncompromising quality standards have made this book an exceptional effort. Their encouragement from time to time has been an inspiration for everyone.

The publisher and the editorial board hope that this book will prove to be a valuable piece of knowledge for researchers, students, practitioners and scholars across the globe.

List of Contributors

Tobias M. Ntuli
Plant Germplasm Conservation Research, School of Biological and Conservation Sciences, University of KwaZulu-Natal, Durban, Department of Life and Consumer Sciences, University of South Africa, Florida, Johannesburg, South Africa

Gustavo Gabriel Striker
IFEVA-CONICET, Faculty of Agronomy, University of Buenos Aires, Argentina

Conceição Santos and Eleazar Rodriguez
Laboratory of Biotechnology and Cytometry, Department of Biology & CESAM, University Aveiro, Aveiro, Portugal

A. Ramesh Sundar, E. Leonard Barnabas, P. Malathi and R. Viswanathan
Plant Pathology section, Sugarcane Breeding Institute (ICAR), Coimbatore, Tamil Nadu, India

Hasnaâ Harrak
National Institute of Agricultural Research, Marrakesh, Morocco

Marc Lebrun and Samira Sarter
International Cooperation Centre of Agricultural Research for Development, Montpellier, France

Moulay Mustapha Ismaïli Alaoui and Allal Hamouda
Hassan II Agricultural and Veterinary Institute, Rabat, Morocco

Thomas Sawidis
Department of Botany, University of Thessaloniki, Thessaloniki, Macedonia, Greece

Nuran Ekici
Department of Science Education, Faculty of Education, University of Trakya, Edirne, Turkey

Feruzan Dane
Department of Biology, Faculty of Sciences, University of Trakya, Edirne, Turkey

Wang Jun and Xia Nian-He
Key Laboratory of Plant Resources Conservation and Sustainable Utilization, South China Botanical Garden, Chinese Academy of Sciences, Guangzhou, China

Wang Jun
Institute of Tropical Biosciences and Biotechnology, Chinese Academy of Tropical Agricultural Sciences, Haikou, China
Graduate University of the Chinese Academy of Sciences, Beijing, China

Ling Fan, Wen-Ran Hu, Yang Yang and Bo Li
Institute of Nuclear and Biological Technologies, Xinjiang Academy of Agricultural Sciences, Nanchang Road, Urumqi, China

Takahiro Tezuka
Graduate School of Life and Environmental Sciences, Osaka Prefecture University, Japan